Time in Roman Religion

Routledge Studies in Ancient History

1 **Cicero and the Catilinarian Conspiracy**
Charles Matson Odahl

2 **Ancient Graffiti in Context**
Edited by Jennifer Baird and Claire Taylor

3 **Roman Elections in the Age of Cicero**
Rachel Feig Vishnia

4 **Time in Roman Religion**
One Thousand Years of Religious History
Gary Forsythe

Time in Roman Religion

One Thousand Years of Religious History

Gary Forsythe

Routledge
Taylor & Francis Group
NEW YORK LONDON

First published 2012
by Routledge
711 Third Avenue, New York, NY 10017

Simultaneously published in the UK
by Routledge
2 Park Square, Milton Park, Abingdon, Oxon OX14 4RN

Routledge is an imprint of the Taylor & Francis Group, an informa business

Library of Congress Cataloging-in-Publication Data

Forsythe, Gary.
Time in Roman religion : one thousand years of religious history / by Gary Forsythe.
p. cm. — (Routledge studies in ancient history ; 4)
Includes bibliographical references (p.) and index.
1. Rome—Religion. 2. Calendar, Roman. I. Title.
BL805.F73 2012
292.3'60937—dc23
2011045603

ISBN: 978-0-415-52217-5 (hbk)
ISBN: 978-0-203-11932-7 (ebk)

Typeset in Sabon by IBT Global.

Printed and bound in the United States of America on acid-free paper by IBT Global.

Contents

Abbreviations

In citing ancient sources throughout the text and notes I generally follow those employed in the third edition of *The Oxford Classical Dictionary*, edited by Simon Hornblower and Antony Spawforth (Oxford University Press 1996). In the bibliography at the end of the book I employ the following abbreviations in citing journals.

AJA	*American Journal of Archaeology*
AJAH	*American Journal of Ancient History*
AJP	*American Journal of Philology*
ANRW	*Aufstieg und Niedergang der Römischen Welt*
ARW	*Archiv für Religionswissenschaft*
CJ	*Classical Journal*
CP	*Classical Philology*
CR	*Classical Review*
CQ	*Classical Quarterly*
GRBS	*Greek, Roman and Byzantine Studies*
HSCP	*Harvard Studies in Classical Philology*
HTR	*Harvard Theological Review*
JHS	*Journal of Hellenic Studies*
JRS	*Journal of Roman Studies*
JSS	*Journal of Semitic Studies*
MAAR	*Memoirs of the American Academy in Rome*
MDAI	*Mitteilungen des Deutschen Archäologischen Instituts (Römische Abteilung)*
MEFR	*Mélanges d'Archéologie et d'Histoire de l'École Francaise de Rome*
PBSR	*Papers of the British School at Rome*
PCPS	*Proceedings of the Cambridge Philological Society*
PdP	*La Parola del Passato*
REA	*Revue des Études Anciennes*
REL	*Revue des Études Latines*
RM	*Rheinisches Museum für Philologie*
TAPA	*Transactions of the American Philological Association*
YCS	*Yale Classical Studies*
ZPE	*Zeitschrift für Papyrologie und Epigraphik*

Acknowledgement and Dedication

Words are inadequate in expressing my gratitude to my deceased wife, Dorothy Alice Forsythe. Her love and beauty have been by far the greatest blessing to my life. For thirty years they bestowed upon me the greatest happiness that I have known and shall ever know. Moreover, her constant support and assistance in every conceivable capacity enabled this blind man to achieve his personal dream of becoming a historian of Greece and Rome and to engage in scholarly research. Her presence in my life has truly been my *sine qua non*, and she has fulfilled the meaning of her name by being a divine gift not only in my life, but in those of her relatives and friends.

To the person possessing the most beautiful soul whom I have known, my dearest wife, whose goodness is so rarely equaled and never surpassed. I love thee with all my heart, with all my soul, with all my mind, and with all my might! Indeed, thou art half my soul!

July 29, 2011
Your Most Blessed Birthday

Preface

Ever since I began studying Latin forty years ago in high school, I have found the ancient Roman calendar with all its intricacies to be a most fascinating subject. Over the years opportunities to teach and to conduct scholarly research have enabled me to develop a serious interest in Roman religion and all things calendrical. For many years, however, these interests have been incorporated into my studies of early Roman history, the Roman annalistic and antiquarian traditions, and related matters. The present volume therefore marks a minor departure from my earlier work in that it focuses entirely upon Roman religious history and interrelated temporal concepts. The book is a collection of six essays, each of which is devoted to a specific subject or area of study. Each essay can be read in isolation from the others, but when taken together, they roughly range over the rich and varied one-thousand years of Roman religious history. The first three chapters concern themselves with the calendar and religion of the Roman Republic, whereas the last three primarily focus upon the religious history of the Roman Empire. I have translated all Latin passages into English, and the rare Greek word or phrase has been transliterated into the Roman alphabet. This policy has been adopted so as to make the material contained in this volume as accessible as possible to students of ancient history and classical studies, as well as to scholars of religious history who happen to be neither ancient historians nor classicists.

The principal analytical tool employed throughout this volume is time. The essays assembled in this book examine how Roman religious thought and behavior were structured into temporal concepts and especially into the ceremonies of the official Roman yearly calendar, one of our most important sources of information about ancient Roman religious history. These six studies are unified by the important role played by various concepts of time in Roman religious thought and practice. Modern studies of archaic Roman religion have discussed how the placement of religious ceremonies in the calendar was determined by their relevance to agricultural or military patterns of early Roman life. But in my opinion modern scholars have failed to recognize that many aspects of Roman religious thought and behavior in later historical times were also preconditioned or even substantially influenced by concepts of time basic to earlier Roman religious history.

Like all societies, the Romans from an early date charted the passage of time by means of a calendar and used the latter as an important framework into which they structured their activities, both public and private. Indeed, the calendar of a society is an important cultural artifact, well worth the attention of historical study, especially in a poorly documented field such as ancient Greek and Roman history. As a written document, a calendar, bearing brief notations of the weekly cycle, anniversaries, festivals, and holidays, represents a skeletal outline of a people's history; and the conscious or deliberate organization, reform, and restructuring of time in the form of a calendar or similar schedule of regularly occurring events can play an important role in shaping a society's complex self identity. To take but a single example, in the modern nation of Israel the season of spring is punctuated with a series of important national and religious observances: Pesach or Passover, traditionally believed to commemorate the Hebrews' exodus from Egypt (beginning 15 Nissan = April 20, 2008); Yom Hashoah or Holocaust Remembrance Day (Nissan 27 = May 2, 2008); Yom Hazikaron or National Memorial Day to honor those who have died in defense of the modern nation of Israel (Iyar 4 = May 7, 2008); Yom Haatzma-ut or Israel's Independence Day, marking the creation of the modern state of Israel in 1948 (Iyar 3 = May 8, 2008); and Shavuot, traditionally believed to be the day on which Moses received the Law on Mount Sinai (Sivan 6 = June 9, 2008). These five holidays form an interesting pattern: two ancient religious festivals coming at the very beginning and end of the series and bracketing the other three holidays of modern creation designed to honor the victims of the Holocaust, Israeli war dead, and the founding of the state of Israel. This pattern, whereby hallowed religious festivals are combined with more recent, secular commemorations are structured into an official calendar of events, so that their aggregate sum exceeds their individual parts in shaping and expressing a national cultural identity, is likewise characteristic of the ancient Roman official calendar. In fact, important temporal cycles are often so deeply embedded within a society that they are taken for granted by the people whose lives they profoundly influence; and despite their significance, they often form part of the culture's tacit assumptions about itself and therefore go unattested (or poorly so) in the written evidence passed on to posterity.

This volume intends to use how the Romans structured time as a means of elucidating numerous aspects of their collective religious experience. Chapter 1 explores the basic structure of the Pre-Julian calendar and attempts to reveal, *inter alia*, how the placement of festivals and temple dedications can provide us with significant evidence for the Hellenization of official Roman religious thought in early times, as well as how January and February embodied the important religious concepts of firstness and afterness. The latter notion is examined in depth in Chapter 2 in the form of the after days and other days closely related to them. Chapter 3 concerning the enigmatic *argei* represents a case study and uses the temporal placement of the rites in the Roman calendar (beginning on an after day) and their duration (sixty days) as an analytical tool in explaining the ancient evidence surrounding these rites. Chapter 4 concerns itself with a totally different temporal concept, the *saeculum*; and by tracing the origin and history of the *Ludi Saeculares* it serves as a transition between Republican and Imperial times. The study attempts to show how and why this celebration originated and was then repeatedly reinterpreted to conform with the changing needs of Roman society. Chapter 5 samples the fascinating and complex phenomenon of ancient mystery religions by a case study that concentrates upon the cult of Magna Mater and the taurobolium. The relatively abundant epigraphic data with a sizable number of calendrical dates are exploited to try to probe more deeply into ancient religious attitudes and behavior, including the idea of imperial mysteries. Chapter 6 concludes the volume with a wide ranging study of how December 25 became Christmas. In order to understand this process fully, this final essay examines a whole host of interesting and interrelated topics: the ancient evidence for the date of Jesus' birth, Roman Republican attitudes about the equinoxes and solstices, the emergence of solar worship during Imperial times, the Roman adoption of the seven-day week, Constantine's conversion to Christianity, and the conflict between paganism and Christianity during late antiquity.

Finally, it is personally gratifying to have this work go to press one hundred years (a Roman *saeculum*) after the publication of two of the most important modern studies of religion in the Roman world: W. Warde Fowler's *The Religious Experience of the Roman People* and Hermann Usener's *Das Weihnachtsfest*.

1 Preliminary Examination of the Roman Calendar

GENESIS OF THE ROMAN CALENDAR

One of our single most important sources of information on Roman religion and its temporal associations is the Roman calendar; and thanks to the survival of numerous epigraphic versions, it has received much modern scholarly attention.[1] This calendar, however, was the one revised by Julius Caesar (hence, its name "the Julian calendar"), consisting of 365 days with an additional intercalary day inserted every four years. It was established as Rome's new calendar as of January 1, 45 B.C. and continued to be the calendar of the Roman Empire and on through the Middle Ages until revised slightly by Pope Gregory XIII in 1582.

Julius Caesar's reformed calendar replaced a lunisolar one of 355 days with an intercalary month of 22 or 23 days inserted every two years or so; and this lunisolar calendar, which modern scholars usually termed the pre-Julian calendar, had been the Roman state's method of marking the passage of time during the previous four hundred years or so of the Republic. Nevertheless, ancient *testimonia* and the structure of the calendar itself suggest that the pre-Julian system was the third stage in an evolutionary process of early Roman time reckoning. During the last two centuries B.C. Roman antiquarians were of the opinion that Romulus, as part of his role as Rome's first king and city founder, had established a calendar, but one of only ten months; and that his successor, Numa Pompilius, to whom the later ancient tradition ascribed virtually all of Rome's religious institutions, reformed Romulus' scheme by adding two months.[2] Romulus began the year with March with the arrival of spring and had it end with December and the onset of winter. Numa completed the yearly cycle by organizing the dormant time of winter into the months of January and February. Strange as Romulus' calendar may seem, in his world-wide survey of pre-literate peoples and their way of reckoning time Martin Nilsson encountered several similar cases in which a people did not bother assigning names to a season of the year characterized by prolonged heat and aridity or cold; and this notion received additional support and corroboration from Frazer's comparative examples.[3]

According to Macrobius (*Sat.* I.12.3) and Censorinus (20.3), Romulus' ten-month calendar numbered exactly 304 days, and the two ancient writers further specify which of the months had 30, and which had 31 days.

This overly precise scheme is certainly unhistorical and is simply the product of later antiquarian reconstruction. Alternatively, rather than viewing Romulus' calendar as having consisted of ten months of roughly 30 days each to correspond to the lunar cycle, A. K. Michels (1949 330) as offered a different explanation, according to which the ten periods would not have been lunar cycles but periods of varying length and marked by different natural or astronomical phenomena that extended over the entire circuit of the year. A natural calendar of this sort would have been perfectly adequate for the early Roman farmer, and we possess an example of such a calendar in Hesiod's *Works and Days* (ll.383–617). Similarly, in his own treatise on agriculture, Varro (*De Re Rust.* I.27–36) describes a farmer's natural calendar consisting of eight periods. Thus, there are two plausible ways in which we can account for the Roman antiquarian tradition of an initial ten-month (or ten-period) calendar. Furthermore, as suggested by Hesiod and Varro, it could have been used by the early iron-age inhabitants of the site of Rome, scattered among several different hilltop villages, before they had coalesced into an organized community.

The second phase of the Roman calendar is revealed by the three dividing days of the months (kalends, nones, and ides) preserved in both the pre-Julian and Julian calendars: for these days, as the later Romans well understood, had originally marked the appearance of the new moon, the moon's first quarter, and the full moon in a calendar that was based upon the lunar cycle.[4] Indeed, three ancient passages inform us as to how the early Roman lunar calendar operated.[5] A minor pontiff had the duty of observing the night sky for the appearance of a new lunar cycle; and on the day following its observance, the Romans were summoned (*kalatus*, whence *kalendae*, marking the first day of the month) to the curia calabra on the Capitoline, where they were told on what day they were to reconvene in order to hear the *rex sacrorum* announcing the religious festivals, the days available for legal business, and other public activities for that month. The day of this second meeting occurred at the time of the moon's first quarter and was termed the nones, coming nine days (counting inclusively) before the ides, which corresponded to the full moon. Exactly when this lunar calendar came into use among the Romans is uncertain, but since it involved a more sophisticated utilization of astronomical knowledge and also hints at the existence of a self-conscious community with organized public activities, we may plausibly conjecture that the lunar calendar would have been in existence no later than the second half of the seventh century B.C.: for archaeological findings from the Forum suggest that the site of Rome was being organized into an embryonic city-state by the last quarter of that century.[6] Roman adoption of a lunar calendar can be seen as part of the larger pattern of state formation, which, as the result of Phoenician and Greek influence, began to transform central Tyrrhenian Italy during the seventh century B.C. (see Drews 1981).

The general pattern of cultural borrowing by Italian peoples from the more civilized Phoenicians and Greeks is best illustrated by the history of the alphabet. During the first half of the eighth century B.C., the Greeks had adapted the writing system of the Phoenicians and had created the world's first true alphabetic script. By 700 B.C., the Etruscans in turn were borrowing and further modifying the Greek alphabet to form their own distinctive Etruscan script, and the practice of writing was quickly adopted by other native peoples of Italy.[7] The idea of a lunar calendar could have been another cultural attribute introduced into central Tyrrhenian Italy by Greeks and Phoenicians. Furthermore, evidence for interaction between Etruscans and Romans on calendrical matters may exist in their names for the eighth month of the year. If Hoenigswald (1941 204–205) is correct in his analysis of the Etruscan names for the months preserved in a Medieval gloss, *Chosfer*, listed as corresponding to October, derives from the Etruscan word for 'eight' and is therefore exactly analogous to Roman October. As richly documented by Nilsson, months and seasons of pre-literate peoples, among whom there is little, if any, specialized calendrical knowledge or expertise, take their names from commonly known natural phenomena, such as the migration of birds, the breeding season of certain animals, or the observed rising or setting of wellknown constellations. Later, however, when astronomical learning advances and becomes the expert knowledge of a few specialists within a society, the experts are the ones responsible for devising a more accurate calendar; and one hallmark of their handiwork is to dispense with the popular designations of the months and seasons and to replace them with names derived from numbers.[8] Thus, six of the Roman month-names (Quinctilis, Sextilis, September, October, November, and December) may hearken back to the earliest phase of calendrical reform in Rome.

Exactly when the pre-Julian calendar replaced the lunar one is unknown, but in attempting to date this innovation modern scholarly speculation has ranged from the sixth century (generally taken to be the Etruscan phase of Rome's later regal period) to the late fourth century and coinciding with the curule aedileship of Cn. Flavius in 304 B.C.[9] The only bit of solid evidence on this matter is a statement in Cicero's *De Re Publica* (I.25), which informs us that the *Annales Maximi* recorded a solar eclipse as having occurred on the nones of June (June 5) approximately 350 years after Rome's foundation. There was in fact a solar eclipse visible from Rome on June 21 of 400 B.C., but as regards the nature of the early Roman calendar, what is really significant about this datum is that the *Annales Maximi* recorded the eclipse as having been seen on the nones of June. Since a solar eclipse takes place when the moon is positioned between the sun and Earth and thus just before the beginning of a new lunar cycle, a lunar calendar should have recorded this eclipse as having occurred just before the kalends of June. Consequently, dating the event to the nones indicates that the Romans of c.400 B.C. were using the pre-Julian calendar that was divorced from the lunar cycle.

GENERAL STRUCTURE OF THE PRE-JULIAN CALENDAR

The organization of the pre-Julian calendar clearly shows that it was the product of experts who were guided throughout by certain basic principles. The Roman calendrical year originally began with March, as Ovid (*Fasti* III.135–366) makes abundantly clear, and as is also shown by the sequence of months Quinctilis–December that take their names from the numbers 5–10. January and February were initially the eleventh and twelfth months of the year. Macrobius (*Sat.* I.13.5) and Censorinus (20.4) inform us that the framers of the calendar regarded odd numbers as more auspicious than even ones, probably because the latter seemed inherently unstable since they are all divisible by two and can therefore be cut in half as it were.[10] Consequently, all the months except February were assigned an odd number of days, roughly corresponding to the lunar cycle of twenty-nine and a half days: January, April, June, August (Sextilis), September, November, and December were given 29 days, whereas March, May, July (Quinctilis), and October were composed of 31 days. Thus, the pre-Julian calendar comprised 355 days, itself being an odd number. February's even number of days and terminal position in the calendar were designed to express the month's overall character as a somber period devoted to the worship of the dead. In addition, the three dividing days of each month were also assigned to odd-numbered days: the kalends was, of course, on Day 1; the ides fell on Day 13 of the shorter months and on Day 15 of the longer ones; and the nones, which preceded the ides by nine days (counting inclusively as the Romans always did), occurred either on Day 5 or 7. In the epigraphic calendars, these days were always displayed in large letters and abbreviated as KAL, NON, and EID.

Similarly, the epigraphic calendars record a series of 48 religious festivals also written in large letters and abbreviated by the first three letters of the festival's name; and all but four of these ceremonies were assigned to odd numbered days of the month.[11] Three of these four festivals assigned to even-numbered days form a distinctive group whose placement in the month can be explained. These are the Regifugium of February 24 and the two days (March 24 and May 24) marked as "quando rex comitiavit = when the king has officiated in the Comitium." Not only do all these three festivals center around the actions of the *rex sacrorum* in the Comitium, but they fall on the 24th of the month. Since both March and May were composed of 31 days, the 24th came nine days (counted inclusively) before the kalends of the next month, just as the nones always preceded the ides by nine days. Moreover, since the nones in the pre-Julian calendar represented the moon's first quarter in the earlier lunar calendar, we may be certain that

March 24 and May 24 were regarded by the organizers of the pre-Julian calendar as corresponding to the moon's last quarter. Once this association was made, the calendar makers applied the same principle to February by assigning the Regifugium to the 24th of that month. The 24th day's association with the moon is further validated by the notation "Lunae in Graecostasi" found in the *Fasti Pinciani* (Degrassi 1963 48) for the 24th of August. These three words, found in only one of the epigraphic calendars, constitute our sole evidence for the existence of a chapel or small shrine to the moon on or near the Graecostasis, a speaker's platform located at the foot of the Capitoline near the Comitium (see Richardson 1992 182–183). Associating the moon's last quarter with an even numbered day also fits with what we know about Roman superstitions concerning the waxing and waning moon, according to which actions or things needing increase or growth were believed to benefit from the waxing moon, whereas a waning moon was thought to favor activities or things verging upon completion or termination (Taverner 1918). Hence, the nones always fell on an odd numbered day (the 5th or 7th), whereas the even numbered 24th represented the waning moon at the time of the last quarter.

On the other hand, the Equirria of March 14 is quite anomalous, and thus far, no really satisfactory explanation as to its placement on the 14th has been offered by modern scholarship.[12] It is obviously paired with the other Equirria of February 27, the latter representing the ending of the old year and the one on March 14 embodying the arrival of the new year; but the calendar makers could have just as easily assigned the second Equirria to an odd numbered day, such as March 9, 11, or 13. Why they chose the 14th instead, remains a mystery.[13] The only observation offered by this writer is that the two Equirria, which symbolized the conjunction of the old and new years, were separated from one another by exactly two *nundinae*, a *binum numdinum*.

This brings us to another important organizational element of the pre-Julian calendar, the so-called nundinal letters. Just as many peoples today have their work, leisure, and various activities organized in accordance with a seven-day week, so the Romans of the Republic and early Empire had their lives structured around an eight-day week; and this eight-day cycle, termed *nundinae*, is represented in the surviving epigraphic calendars by a continuously running series of letters, A–H.[14] Within each particular locale of the Roman world one specific day of the nundinal cycle was designated the market day, on which all normal daily labor was forbidden to enable farmers and townsfolk to congregate for buying and selling and other sorts of interchange.[15] Each month of the Roman year was so structured that the nones, ides, and the last day of the month shared the same nundinal letter (Michels 1967 88). Thus, the nones and ides were always separated by a nundinal cycle, and the period between the nones and the last day of a month always formed a span of time that the Romans termed a *trinum nundinum*.

HELLENIZATION OF EARLY ROMAN CULTS

A careful examination of temple dedications and other religious activities provide us with important evidence for early Greek influence upon the thinking and religious behavior of Roman priests. These data clearly indicate that Hellenization began very early, even before the close of the regal period, and was quite pervasive throughout the early Republic.

The later Romans believed that one of their oldest cults was that of Hercules at the Ara Maxima in the Forum Boarium, and that it had not only preceded Rome's foundation, but had been established by Hercules himself.[16] There were several peculiarities about this cult that have attracted the attention of modern scholars as to its origin. For example, when sacrifice was made to Hercules at this altar, the person did so in the Greek fashion with the head uncovered as opposed to the normal Roman custom of having the head veiled (Macrob. *Sat.* III.6.17). The practice of successful merchants and generals of offering tithes of their profits and booty at this altar prompted Bayet (1926) to seek a Phoenician origin of the cult; and this hypothesis has been pursued further by other scholars. The ancient sources state that down to the censorship of Ap. Claudius Caecus in 312 B.C., the cult was administered by members of two Roman families: the Pinarii and Potitii. Although the *gens Pinaria* was attested in the consular *fasti* of the fifth century B.C., there is no evidence at all for a *gens Potitia*. Van Berchem (1959–1960) therefore suggested that the Potitii were not members of a Roman family, but were temple slaves, whose existence is well attested in the ancient Near East; and he further proposed that the name Potitii derives from the Latin verb *potior* (to take possession of) and thus must have meant "those possessed (i.e., by Hercules)."

For the purpose of this study, the one element in the cult that is most interesting and significant is the fact that its rites were observed every year on August 12. Given the cult's antiquity and early Roman priests' avoidance of even numbered days for celebrating festivals, this is indeed a curiosity. As a result, Piganiol (1962) postulated that the cult and the day of its celebration were borrowed from Greece, and he found as the Roman model the Kronia observed by the Athenians on the 12th day of Hekatombaion, the first month of the Athenian calendar and roughly corresponding to Roman July. Since the Kronia at Athens was celebrated in honor of Kronos and not Herakles, using the festival to explain the dating of the Roman one does not hold up very well; but Piganiol's basic approach, i.e., seeking a non-Roman origin to explain August 12, makes excellent sense. The Athenians did have a festival in honor of Herakles in the second month of their year, Metageitnion, corresponding to Roman August; but unfortunately, we do not know on what day of the month it was celebrated (see Parke 1977 51–52). In any case, assigning a non-Roman origin to August 12, whether Phoenician or Greek, strikes this author as perfectly plausible.[17]

According to the later ancient Roman tradition (Livy I.45 and Dion. Hal. IV.25–26) King Servius Tullius was responsible for constructing the temple of Diana on the Aventine; and he did so as a means of establishing Roman hegemony over the other Latin communities. Modern scholars have generally supposed that Diana's Aventine temple was in part designed to supplant the famous shrine of Diana at Aricia.[18] Both Livy and Dionysius describe Servius Tullius as having taken the cult of Artemis of Ephesus in western Asia Minor as his model. Although a direct connection between Roman Diana and Ephesian Artemis is unlikely for the sixth century B.C., an indirect one is not and is claimed by Strabo (IV.1.4–5), according to whom the wooden cult statue of the goddess in the Aventine temple was modeled after the one of Artemis revered by the people of Massilia.[19] By the mid-sixth century B.C. the Phocaeans, who had colonized Massilia at the mouth of the Rhone on the southern coast of France, as well as Elea on the Italian coast of Lucania, had emerged as the most commercially active Greeks of the Tyrrhenian Sea (Hdt. I.163–167); and Massiliote interaction with Rome at that time can be assumed. A marble relief discovered at Aricia, dating to the early fifth century and showing Orestes killing Aegisthus, suggests that Latin Diana was by then being equated with Greek Artemis: for the story of Orestes and Aegisthus was part of the larger Greek myth involving Agamemnon's sacrifice of his daughter Iphigenia at Aulis and her miraculous rescue by Artemis.

In 495 B.C., a temple to Mercury was dedicated in Rome on May 15.[20] The temple's day of dedication shows that already in 495 B.C. at least some Roman priests not only equated Roman Mercury with Greek Hermes, but they also were aware of the Greek myth that made Hermes the son of Zeus and an obscure divinity named Maia. Since all ides were sacred to Jupiter (Macrob. *Sat.* I.15.15–18), the Roman equivalent of Zeus, having Mercury's temple dedicated on May 15 was intended to associate Mercury with his supposed father; and choosing the month of May in which to have the shrine consecrated was based upon the erroneous, but Hellenizing notion that the month's name was derived from that of Mercury's mother.[21]

Two years later in 493 B.C. there was dedicated on the Aventine the temple of Ceres, Liber, and Libera (Dion. Hal. VI.94,3 and Tacitus *Ann.* II.49). Ceres was the goddess of increase, especially of agriculture; and the day of the temple's dedication, April 19 that became the annual festival of the Cerialia, must have been deliberately chosen by Roman priests, because it occurs amid an important series of spring festivals: the Fordicidia of April 15, the Parilia of the 21st, the Vinalia of the 23rd, and the Robigalia of the 25th. Although all three members of this Aventine triad were native to Italy (Radke 1979 88–91 and 175–183), at the time of the temple's dedication they were already equated with Greek Demeter, Iacchus, and Persephone, who were widely worshipped among the Greeks of Sicily and southern Italy (Diod. V.4–5). According to Dionysius (VI.17.2–4) the temple was vowed after consultation of the Sibylline Books, a collection of enigmatic writings composed in Greek verse that the Romans had probably obtained recently from the Greek city of Cumae (Parke 1988 85–89), located south of Rome on the Italian coast a mere 115 miles from the mouth of the Tiber. Pliny (XXXV.154) informs us that the temple was decorated by two Greek craftsmen, Gorgasus and Damophilus; and Greek rites of the temple were conducted by a priestess brought in from Magna Graecia (Cic. *Pro Balbo* 55).

The religious history of the early Republic contains three other events that involve the importation of purely Greek cults and rites for which the Romans had no equivalents. In 484 B.C., the Romans dedicated in the Forum a temple to the Greek divine twins Castor and Pollux (Livy II.42.5); and the plausibility of such an early date for this cult in Rome is strengthened by the discovery at Lavinium of a dedication to these Greek gods dating to c.500 B.C. (Gordon 1983 #2). According to Livy (IV.25.3 and 29.7) an epidemic in Rome in 433 B.C. led to the consultation of the Sibylline Books and the dedication two years later of a temple to the Greek god Apollo, whose name the Romans adopted into Latin without any change. For the year 399 B.C. both Livy (V.13.4–8) and Dionysius (XII.9) record the outbreak of another plague and the consultation of the Sibylline Books, which resulted in the Romans celebrating for the first time a ceremony that they termed a *lectisternium* (couch spreading).[22] The rite was simply the Roman adoption of the Greek *theoxenia*, a public religious banquet with images of gods placed upon couches before which were set tables decked out with food. Livy and Dionysius agree in stating that the divinities so honored by the Romans on this occasion were Apollo, Latona, Diana, Hercules, Mercury, and Neptune. Dionysius further indicates that the six deities were grouped in three pairs in the order just stated. The pairing of Diana and Hercules is quite anomalous. There was nothing significant in Greek cult or mythology that joined Artemis with Herakles. Their association in Rome in 399 B.C. might have been due to the Roman desire in time of plague to enlist the two divinities' well-known concern respectively for women in childbirth and manly vigor; and this coupling of the two deities might have also been suggested to Roman priests by the juxtaposition of their major festivals in the calendar on two consecutive days: August 12 and 13. From the standpoint of Hellenization, the list contains two noteworthy features. Leto, whom the Romans rendered into Latin as Latona, was important to the Greeks only as the mother of Apollo and Artemis; and her inclusion in this ceremony is indicative of direct Roman borrowing. Although Neptune had originated as a divinity of fresh water, especially of springs and streams for the Roman peasant farmer (see Gell. XIII.23.2), he was here obviously being identified with Greek Poseidon, the mighty god of the sea: for his pairing with Mercury must have been intended to ensure trade by sea, especially the importation of grain.

OVERVIEW OF THE YEARLY FESTIVALS

A survey of the Roman official religious calendar will serve to elucidate further Roman religious thinking and behavior, especially of the priests, which are important in forming the underlying methodological and heuristic premises of the other chapters of this book.[23] Unfortunately, our knowledge of these festivals is rather uneven. It is more detailed for the festivals celebrated during the months of January to June, thanks to the survival of Ovid's *Fasti*, which covers only these six months.[24]

The earliest festivals of the religious calendar clearly indicate that archaic Roman religion was largely the religion of the Roman peasant farmer, whose survival depended upon the productivity of his crops and animals. In some instances, farmers performed religious rites on their own land to enlist divine assistance or to ward off harm. In other cases, the religious ceremonies were conducted by Roman priestly officials who were acting on behalf of the entire community. They thus insured that the rituals were performed correctly so as to obtain the desired favor of the gods. Besides agriculture and stock breeding, the calendar also reflects two other concerns of the Roman community: success in war against foreign threats and the overall health and reproductive viability of all individual Romans. In most instances the priests' assignment of the festivals to specific days of the year can be understood in terms of the natural agricultural cycle of Latium; but an examination of the festivals and their placement in time can also reveal in what ways and to what extent Roman priests applied temporal placement more abstractly to symbolize and embody the human life cycle from birth to death.

As already mentioned, the Roman year originally began with March, *mensis Martius*, taking its name from the important Italic god Mars (Ovid *Fasti* III.9–98). In order to symbolize the beginning of a new year, on March 1, as Ovid further informs us (*Fasti* III.137–150), the Regia and the houses of the flamens were festooned with fresh laurel, and the sacred fire of Vesta was extinguished and rekindled. The midpoint of the month (March 15) was given over to the celebration of the return of spring as the Romans enjoyed themselves outdoors to honor Anna Perenna, a minor divinity embodying the recurring cycle of the year (Ovid *Fasti* III.523–542).[25] During the month, Rome's leaping priests, the *salii*, organized into two groups of twelve, performed their sacred dance throughout the city. As they danced, they struck their sacred shields known as *ancilia* with a staff or short spear and sang a song, the carmen saliare, whose Latin was so old that later Romans had no clear idea as to what it meant.[26] The purpose of these rites was probably twofold: agricultural and military (Fowler 1899 39–41 and Frazer 1929 III.61–68). Their leaping dance was most likely intended to promote the growth of crops through sympathetic magic, and the noise created by the banging on their shields must have been designed to drive off evil spirits that might endanger the Romans, their crops, and their herds. Later Roman antiquarians detected in their song reference to mamurius Veturius, who was conjectured to have been the craftsman of the shields; but the phrase is probably an archaic rendering of *Marmar Vetus*, Old Mars, the embodiment of the past year that must now give way to the new one with the arrival of March.[27] Moreover, since the salii are known to have performed their dance on October 19, the Armilustrium that signified the ending of the military campaigning season, their dancing on March 19 is likely to have been intended to mark the opening of the same. In addition, the Equirria of March 14 and the Liberalia of the 17th may have originally served a military purpose in symbolizing the need to prepare horses for cavalry service and to make the newest crop of Roman male teen-agers available for service in the army: for on the Liberalia it was customary for Roman boys in their mid teens to formally lay aside the *toga praetexta* and *bulla* of childhood and to assume the *toga virilis* of manhood (Ovid *Fasti* III.771–790 and Cic. *ad Att.* VI.1.12), at which time the youths were eligible to serve in the army. Finally, the theme of the human life cycle is apparent in the Roman priests assigning to the first day of the year, March 1, the dedication of a temple devoted to childbirth (Festus 131L s.v. *Martias Kalendas*). The shrine, located on the Esquiline, was consecrated to Juno, to whom all kalends were sacred (Ovid *Fasti* I.55 and Macrob. *Sat.* I.15.18). The sanctuary was dedicated in 375 B.C. (Pliny XVI.235), and Juno received the title Lucina from the sacred grove (*lucus*) in which it was located; but in later times Lucina was reinterpreted as deriving from *lux, lucis* (light), because Juno Lucina brought babies from their mothers' wombs into the light of day (Ovid *Fasti* III.245–258, Plut. *Quaest. Rom.* 77, and Macrob. *Sat.* VII.16.27–28).

Later Romans developed two different explanations for the derivation of April's name. One notion was that it came from Aphrodite, and the other was that it derived from the verb *aperio* (to open), because it was the month of blossoming plants.[28] Modern scholars have likewise been divided as to the origin of April. For example, linguists such as Whatmough (1932 158–159) and Hoenigswald (1941 199) derive it from Etruscan *Ampile* or *Amphile*, preserved in a Medieval gloss, and which they in turn conjecture to have come from Greek *Aphrios*, a month attested in Thessaly (see Samuel 1972 86 and Bickerman 1980 20). On the other hand, Radke (1993 129) has suggested an Indo-European etymology for the month's name by connecting it with Sanskrit *áparah*, corresponding semantically to Latin *alter*, a comparative form meaning 'the other of two', so that April would have originally meant the other or second month in comparison to March. Students of Roman religion such as Fowler (1899 66) have not hesitated to accept Varro's derivation from *aperio*, because, as observed by Varro himself, whose opinion was opposed to that of Fulvius Nobilior and Junius Gracchanus, there was no trace of Aphrodite to be found in early Roman writings. Nevertheless, the month's association with Aphrodite / Venus resulted in the dedication of a temple to Venus Verticordia in 114 B.C. on April 1 (Obsequens 37 and Oros. V.15.21–22).

In any case, whatever the derivation of the month's name (and the author inclines toward *aperio*), there is no doubt that it was in fact the time of blossoming plants; and the major Roman religious festivals clearly reflect the early Romans' preoccupation with protecting their crops and herd animals. At the Fordicidia of April 15 the pontiffs sacrificed thirty pregnant cows, one for each of the thirty *curiae* of the Roman people. The offering was made to Tellus (Earth), and the unborn calves were removed from the sacrificial victims, their bodies were burned to ashes, and the latter were preserved by the eldest Vestal for use six days later at the Parilia (Ovid *Fasti* IV.630–640). In describing the origin of this ceremony Ovid (Fasti IV.641–672) tells how Numa once averted the infertility of the land and of the herd animals by seeking advice from Faunus in a dream. The fertility of the sacrificed cattle was designed to ensure the productivity of the land and of its animals; and the ceremony was placed in the calendar just before three other festivals for securing the safety of cattle (the Parilia of April 21), the health of the grapes on the vine (the Vinalia on the 23rd), and the protection of the grain from disease (the Robigalia on the 25th). Among these doubtless primordial rites the Roman priests later inserted the Cerialia by having the Aventine temple of Ceres, Liber, and Libera dedicated on April 19; and by the second century B.C. April ended with the Ludi Florales, devoted to Flora, the goddess of flowering plants (Ovid Fasti IV.943–947, V.275–294, 312–330, and 371–376).

Despite Roman antiquarians' desire to derive May's name (*mensis Maius*) from excogitated minor divinities such as Maia or Maiestas, the name clearly derives from a basic Italic *etymon mai-*, meaning 'big', because the month was characterized by the growth of the farmer's crops.[29] In fact, the time was so critical for farmers that the overall nature of the month was somber, so much so that it was regarded as inauspicious to marry at this time (Ovid *Fasti* V.485–492 and Plut. *Quaest. Rom.* 86). In order to avert untoward mischief from the spirit world, the Romans performed simple nocturnal rites in their own homes to appease ghosts on the Lemuria of May 9, 11, and 13 (Ovid *Fasti* V.419–444). This "festival" was the only one in the Roman calendar that was assigned to three successive odd-numbered days, doubtless because the Romans regarded the thrice repetition of the rites as particularly efficacious and necessary. The Lemuria was then followed on May 15 by what Plutarch (*Quaest. Rom.* 86) called the greatest of purifications, involving the disposal of the *Argei*, which will be treated more fully in the third chapter of this work. The kalends of the month, rather than being devoted to the worship of Juno, was sacred to another female divinity, Bona Dea, a rather enigmatic goddess of fertility, from whose temple and rites all males were excluded.[30] The same day was also sacred to the Lares Praestites, the Guardian or Protecting lares (Ovid *Fasti* V.129–146). They formed an integral part of every Roman household, in which they received worship at a small chapel and were thought to safeguard family, home, paths, streets, and farmland. Archaeological excavations, especially at Pompeii, have uncovered numerous figurines and reliefs of these divinities from household shrines and those set up at crossroads. The Lares were usually depicted as two males sometimes with a dog lying or sitting at their feet.[31] The latter was obviously used to epitomize their vigilant and protective nature (Plut. *Quaest. Rom.* 51). Roman reliance upon them to protect crops from harm is demonstrated by their invocation at the very beginning of the Arval Hymn (*ILLRP* 4 = Gordon 1983 #75): "Oh help us, ye Lares! Oh help us, ye Lares! Oh help us, ye Lares! = E nos, Lases, iuvate! E nos, Lases, iuvate! E nos, Lases, iuvate!" The month ended with the Ambarvalia, the circuit of the fields, in which farmers sacrificed a pig, calf, and lamb (the *suovetaurilia*) and used their entrails to perform a purification of their farmland from all possible harm.[32]

In Roman antiquarian literature three different explanations were given for the name of June: (1) from *iuniores*; (2) from Juventas; (3) from juno. According to Censorinus (22.9) Fulvius Nobilior and Junius Gracchanus thought that May and June took their names respectively from *maiores* (the elders) and *iuniores* (the younger ones, i.e., those fit for military service).[33] The second explanation involving the minor divinity Juventas (Youth = Greek Hebe) is found only in Ovid (*Fasti* VI.65–82). According to Macrobius (*Sat.* I.12.30) the third explanation was offered by Cincius, who observed that a month called Junonius was to be found at Aricia and Praeneste; and in recording the same derivation Ovid (*Fasti* VI.57–64) adds to this list the Latin towns of Lanuvium, Lavinium, and Tibur. There can be little doubt that this third explanation of the month's name is correct. In early times Juno was the embodiment of youthful vigor, both male and female; and only later was she more narrowly associated with women (Palmer 1974 3ff). Moreover, since the crops in Latium would have been attaining their maturity during this month, naming it after Juno would have been perfectly natural and also shows how early Roman religious thought easily associated a human attribute such as youthful vigor with a similar quality exhibited by plants. In 344 B.C. the temple of Juno Moneta on the Arx was dedicated on June 1 (Livy VII.28.4–6, Ovid *Fasti* VI.183–190, and Macrob. *Sat.* I.12.30); and since, as already noted, all kalends were sacred to Juno, this day could have been viewed as being the most Junonian day in the entire year. The day was also sacred to a minor divinity called Carna, who watched over one's internal organs; and the day was characterized by offerings of bean-meal and lard to the goddess (Ovid *Fasti* VI.169–182 and Macrob. *Sat.* I.12.31–33). Thus, the day was a kind of "physical fitness day." This theme was further structured into the calendar with the Matralia of June 11 in honor of Mater Matuta, a goddess of ripening and maturation. Not only was her worship timed to coincide with the maturation of the crops, but was also connected with human maturation, especially of young females. Both Ovid (*Fasti* VI.559–5562) and Plutarch (*Quaest. Rom.* 17)inform us that in worshipping Mater Matuta on this day Roman mothers did not pray for the welfare of their own children, but for those of their sisters. H. J. Rose (1934 156–157) ingeniously explicated this peculiar custom through etymology by suggesting that rather than praying for the welfare of the children of their sisters (*sorores*), the mothers originally sought divine assistance for their own swelling (*sororii*) children, i.e., those entering upon puberty.[34] Finally, the ceremonial cleaning of Vesta's temple on June 15 symbolized the community's readiness for taking in the summer harvest (Ovid *Fasti* VI.713–714).

Once we reach the end of Book VI of Ovid's *Fasti*, we are confronted with six months whose festivals are very poorly documented. Apart from brief statements preserved in writers such as Festus or in Varro's terse overview of the festivals in VI.12–26 of his *De Lingua Latina*, we are often left with little more than the names of the festivals themselves abbreviated in the epigraphic calendars with only our conjectures to guide us. Unlike the four months just surveyed, whose very names reflect their close association with the agriculttural cycle, the next six months simply took their names from numbers. Yet despite the exiguous surviving evidence, their festivals were clearly related to the heat and harvest of summer followed by the autumn planting. July 7, the Caprotine Nones, commemorated the pollination of fig trees during midsummer (Forsythe 1994 322–324). The Neptunalia of July 23 honored the early Roman god of springs and streams at a time when fresh water was at a premium. In later times Neptune was equated with the Greek Poseidon and thereby was changed from a god of fresh water worshipped by the early Roman peasant into a mighty god of the sea. Vulcan, the Roman god of destructive fire, received worship on August 23 when the dry heat of summer must have posed a serious threat of combustion (Rose 1933). The end of the summer harvest was marked by three festivals: the Portunalia of August 17, the Consualia of August 21, and the Opiconsivia of August 25. Contrary to what his name might suggest (see Fowler 1899 202–203), Portunus was probably not originally a god of harbors (*portus*), but one who looked after gates (*portae*) and doorways: for Festus (262L s.v. *Portum*) indicates that *portus* in the Law of the Twelve Tables meant the door of a house. It has therefore been plausibly conjectured that Portunus had his festival placed at this point in the calendar because of farmers needing to open and close their barns and silos in storing their harvested grain. Similarly, Consus was the god of storage, and Opiconsivia was simply 'the abundance of Consus', a most important attribute for farming families, whose survival depended upon the quantity of their harvest. As her name suggests, Opiconsivia probably originated as a mere attribute of Consus before evolving into a minor divinity in her own right; but given the great importance that agricultural productivity obviously had in early Rome, the priests prudently retained Opiconsivia's close association with Consus by the placement of her festival in the calendar.

During the autumn months the Roman farmer was busy with plowing and planting next year's crops as well as harvesting grapes to be processed into wine. The end of the autumn vintage was marked in the religious calendar by the festival of the Meditrinalia of October 11. Similarly, the autumn planting of wheat and barley was represented in the rites of October 5 and November 8 in which a ceremonial storage pit called the Mundus was opened, signalling the removal of seed grain to be planted. This religious ceremony, intended to confer ritualistic correctness and thus divine favor upon this important agricultural activity, was later reinterpreted to mean that on these days the doorway to the Underworld was opened and ghosts were about (Fowler 1912). The end of the military campaigning season with the onset of colder and rainy weather was marked by the Armilustrium of October 19 in which men assembled under arms at the Armilustrum on the Aventine outside the pomerium and underwent ritualistic purification from the pollution of bloodshed.[35] This ceremonial closing of the campaigning season came exactly seven months after its formal opening by the salii on March 19. As in the case of the summer harvest, the close of the autumnal planting season was represented by three festivals in mid December: a second Consualia on the 15th, the Saturnalia on the 17th, and the Opalia on the 19th. Saturn was originally the Roman god of sowing, and like Consus the god of storing grain, he was given the attribute of abundance (*ops*), so that the farmer's planting would result in an abundant yield.

January and February form the most curious pair of months in the calendar. Although they were originally the last two months in the year, at some unknown date the Romans shifted the beginning of their year from March 1 to January 1, so that January and February became the first two months of the year. Moreover, since Books I–II of Ovid's *Fasti* are devoted to these two months, their festivals are fairly well attested; but before examining some of these festivals for what they can reveal concerning things temporal in Roman religious thinking, we must first consider the two months themselves.

As already noted, the later Romans regarded these two months as unusual in that they had not been part of Rome's earliest calendar attributed to Romulus, but they were later added by Numa. Their peculiar character and obvious pairing together are further suggested by the etymology of their names and by the fact that they both end in *-uarius*. Unlike the preceding six months whose names derive from numbers, these two months' names resemble the other four months in the calendar in having names that represent their overall character. As the very last month in the year, February was largely given over to rites to honor the dead; and the ancient sources are unanimous in deriving February from *februum*, a term used to describe things used in rites of purification.[36] The month's somber nature was further underscored by the fact that it was the only month given an even number of days. January, on the other hand, took its name from Janus, the god of doorways, entrances, comings and goings, and all manner of passageways.[37] As a result, in the more abstract thinking of Roman priests it was necessary to invoke Janus first in prayers in order to be granted access to the other gods.[38] Thus, these two months must have constituted to early Romans an obvious juxtaposition and antithesis in which this period of the year, coming between the end of the autumnal planting and the new growing cycle of spring, was organized into two distinct parts: January forming the beginning of the period, and February bringing it to an end.

we can say that January embodied priority and February posteriority; or to use more ordinary words, they possessed the respective qualities of firstness and afterness. This antithesis is well illustrated more simply in the calendrical juxtaposition and conceptual bipolarity of the dividing days and the *dies postriduani* (or "after days"), the latter coming immediately after the former. This subject will be explored in detail in the next chapter, but it is most interesting and relevant for this study to see the same calendrical juxtaposition and conceptual bipolarity exhibited on a larger scale in terms of two entire months.

January 1 was sacred to Aesculapius, whose name was the Roman rendering of the Greek god Asclepius. As the result of a plague in Rome in 293 B.C., the Romans consulted the Sibylline Books, and on their advice they sent ambassadors to Epidaurus in mainland Greece to bring to Rome the famous cult of Asclepius.[39] As the story goes, the god in the form of a snake was transported on shipboard back to Rome, where his temple was dedicated on January 1 of 291 B.C. It was located on the southern end of the Tiber Island on the site now occupied by the Church of San Bartolomeo. In the course of time the god's temple became a sacred house of healing involving rites of incubation. People seeking cures prayed and slept in the temple in hope that they would wake up healed or would be told by the god in a dream what remedy or regimen was needed to heal their ailment. Numerous dedications, some testifying to miraculous cures, have been found in the bed of the Tiber near the island.[40] An interesting question arises from the temple's day of dedication. In later times it coincided with the first day of the Roman year, but was this the case in 291 B.C.? Due to the outbreak of a serious war in Spain in 153 B.C. the Romans changed the day on which the consuls entered office from March 15 to January 1 (Livy *Per.* 47). Henceforth the consular year always began on January 1, but it seems likely that this day had long supplanted March 1 in marking the commencement of the Roman year (see Michels 1967 97–100). If this were the case in 291 B.C. as is quite possible, January 1 as the day of dedication for Aesculapius' temple might have been deliberately chosen so as to begin each Roman year with the worship of the god responsible for ensuring people's health.

The only other festival of January requiring examination for this study is the Carmentalia. No other festival in the Roman calendar shared its temporal characteristics. It was unique in that there were two days assigned to the worship of Carmenta: the 11th and 15th, so that they fell on the odd numbered days on either side of the ides sacred to Jupiter. Carmenta was a divinity of childbirth and prophecy. Her two traits were doubtless intertwined logically, because in addition to asking for assistance during childbirth itself, women are likely to have sought out the advice of a prophetic divinity to learn whether they had conceived, when the child would be born, whether it would be healthy and would survive infancy, what its sex was, would there be complications with the delivery, etc. In addition,

Carmenta had two epithets: Prorsa or Porrima and Postverta (Ovid *Fasti* I.633–636 and Gell. XVI.16). Ovid engages in poetic fancy in explaining the terms as the names of Carmenta's sisters or fellow fugitives from Arcadia, and that Porrima had the power of foretelling the future, whereas Postverta could see into the past. Gellius, on the other hand, cites and follows Varro, who correctly understood the names as referring to infants being born forward (head first) vs. those born backwards (feet first). The epithets in turn could account for the peculiar placement of the Carmentalia in the calendar: for just as there were two major aspects of Carmenta, so her worship was officially organized not into one, but two days; and furthermore, Carmenta Prorsa or Porrima was honored two days before the ides of the month, whereas Carmenta Postverta received worship two days after the ides. Lastly, placing the festival of this divinity of childbirth in the middle of the month that took its name from the god of beginnings once again demonstrates how early Roman priests symbolically incorporated the human life cycle into the yearly one.

In a single elegiac couplet Ovid (*Fasti* II.67–68) informs us that on February 1 a sacred grove near the Tiber received worship: "Then also is celebrated the sacred grove of Alernus, next to where the sojourning Tiber seeks the sea waters = Tum quoque vicini lucus celebratur Alerni, | Qua petit aequoreas advena Thybris aquas." The best reading of the surviving manuscripts of Ovid's *Fasti* is *Alerni* for the name of the divinity; but some texts instead read *Averni*. Yet, since neither Alernus nor Avernus are otherwise attested in Roman cult, modern scholars have not been satisfied with either reading. *Helerni* is the most commonly accepted emendation printed in modern editions of the *Fasti*; but Helernus is also otherwise unattested and is the excogitation of modern scholars of Roman religion (e.g., Wissowa 1912 236), who derive his name from *holus, holeris* (= vegetable) and make him out to be a god of vegetables. A simpler and far more satisfying solution to the problem can be achieved by changing a single letter, emending *Alerni* to *Aterni*. The latter divinity is mentioned in a brief entry in Paulus Diaconus' epitome of Festus (83L s.v. *furvum*), in which the archaic adjective *furvus*, meaning 'black', is explained: "They used to sacrifice to Aternus a *furvus* (that is, black) ox = Furvum bovem, id est nigrum, immolabant A[e]terno." Although the text of Festus reads *Aeterno*, modern scholars have generally regarded it as a mistake for *Aterno*, because the context clearly indicates that the sacrifice was made to a divinity of darkness, not of everlasting time. As will be discussed further in Chapter 2, the adjective *ater* (black) was commonly applied to a day that came immediately after a dividing day, so that *dies postriduani* and *dies atri* were synonyms used interchangeably. In fact, in this context *ater* did not originally mean 'black', but 'other' or 'next', because this use of *ater* was a dialectal variant of *alter*, as seen in Quinquatrus, the name for the festival on March 19, so called because it was the "fifth-next" day after the ides. Thus, the original meaning of *dies ater* was not 'black day', but 'next day', i.e., the day after a dividing day; but at some relatively early date the inauspicious character of all *dies atri* caused them to be reinterpreted as black days. Consequently, Aternus must have originally been the divine embodiment of afterness or otherness and along with the *dies atri* was reinterpreted to be the god of darkness and other-worldliness, i.e., a god of the netherworld. As argued in the preceding paragraph, January and February formed an antithetical pair in the Roman calendar, representing the contrasting concepts of firstness and afterness. It therefore would have been perfectly natural for the early Romans to have consecrated the first day of February to Aternus, the god of afterness. Additional support for this reconstruction is found in a lead curse tablet from Rome and dating to the late Republic (*ROL* IV. 280–285). The author of the text, who sought the bodily torture and death of a personal enemy, accompanied his plea to Proserpina Salvia with the sacrifice of a black pig and hoped to add further efficacy to his prayer by beseeching that his enemy's demise be accomplished during the month of February, the month most closely associated with the dead.

Finally, we shall conclude this overview of the Roman calendar quite appropriately with an examination of the Terminalia and Regifugium of February 23 and 24. As in the case of some other festivals treated in the preceding paragraphs, early Roman priests placed the Terminalia in the calendar so as to serve a dual purpose. The festival was, of course, intended to reinforce the sanctity of boundary stones used to mark off farmland (Ovid *Fasti* II.639–684), but if that had been the only purpose behind the festival, early Roman priests could have assigned it to virtually any time of the year. Their decision to place it toward the end of February, originally the last month of the year, must have been designed to serve a second, symbolic purpose of marking off the termination of the yearly cycle itself. This function of the festival is clearly borne out by its coupling with the Regifugium that followed it on the very next day. As already demonstrated, the 24th day of a month represented to the organizers of the calendar the waning of the moon and hence of the month itself. Assigning the Regifugium to February 24 was obviously intended to signify the waning of the entire year. Plutarch (*Quaest. Rom.* 63) informs us (doubtless in reference to the Regifugium) that after performing a sacrifice in the Comitium, the rex sacrorum fled from the scene. In this matter the king of sacrifices personified and embodied the Roman state, and his flight represented the passing away of the Roman year. Moreover, in his study of this festival E. T. Merrill (1924 37–38) made the ingenious suggestion that the five-day *interregnum* used in Republican times to bridge temporal gaps between consulships had its origin in the five-day interval between the Regifugium and March 1, thus making this period of time the primordial *interregnum* of the Roman state.[41] Whenever the Romans needed to use intercalation to bring the calendar back into closer alignment with the solar year, they always placed the intercalary month (and later, following the Julian reform, the single intercalary day) between the Terminalia and the Regifugium (Censorinus 20.6 and Macrob. *Sat.* I.13.15). By doing so the pontiffs caused the Terminalia to be the last day of February, the last month of the year except when dislodged through intercalation; and they further ensured that the year came to a close with the five-day *interregnum* between the Regifugium and March 1.

2 The After Days and Other Curiosities

In the summer of 101 B.C. two great armies joined battle at Vercellae (modern Vercelli) near the foot of the Alps in northwestern Italy about 42 miles west of Milan. One army consisted of the Cimbri, a Germanic tribe who for the past few years along with their kindred Teutones had been raiding and terrorizing parts of Gaul and Spain in what the Romans termed the Germanic War. The other army at Vercellae was Roman and was commanded jointly by C. Marius and Q. Lutatius Catulus. By bringing the war against Jugurtha in North Africa to a successful conclusion in 105 B.C., Marius had established himself as Rome's most capable general of the day. In the same year on October 6, the Romans had suffered a crushing defeat at Arausio (modern Orange) in southern Gaul; and in order to try to reverse what seemed like the inevitable advance of the Germans upon Italy itself, the Romans elected Marius to consecutive consulships and placed him in charge of the Germanic war. At Vercellae, Marius was holding his fifth consulship, whereas Catulus, having been consul in the previous year, was now serving as proconsul and was a man of much less military experience. The outcome of the battle was a decisive victory for the Romans and ended the Germanic threat. In his account of the battle, Plutarch (*Marius* 25–27 with 26.2+4–5) says that the engagement was fought on July 30, and during it Catulus vowed a temple to the Fortune of that day. Both the *Fasti Pinciani* and the *Fasti Allifani* (Degrassi 1963 47 and 179) indicate that July 30 was the day of dedication for a temple to Fortuna Huiusce Diei in the Campus Martius. It is therefore obvious that Catulus had the temple dedicated on the anniversary of Vercellae. Thanks to a brief description of this shrine by Varro in his *De Re Rustica* III.5.12, the structure has been generally recognized as Temple B in the Area sacra di Largo Argentina in the southern Campus Martius (Richardson 1992 33–35 and 156).

In *De Legibus* II.28 Cicero lists "the Fortune of this very day (or more simply, 'Today's Fortune')" among the cults to be included in his ideal state; and he justifies the inclusion of Fortuna Huiusce Diei by observing, "she prevails with respect to all days = nam valet in omnis dies." Today's Fortune neatly epitomizes an important aspect of Roman religious thought discernible from earliest times and enduring to the very end of Roman civilization,

Figure 2.1 Temple B of the Area Sacra di Largo Argentina in the Campus Martius in Rome (Author's Photograph).

namely, the idea that each day of the year had its own character, and that one crucial duty of the pontiffs was to be well informed in these matters so as to safeguard the Roman community by assigning all major public and private activities to their appropriate temporal settings. The dedication dates of temples carefully enshrined in the Roman calendar from the early years of the Republic clearly demonstrate that Roman priests were already then acting in accordance with these ideas. We may even plausibly surmise that such calendrical conscientiousness extended back into the regal period: for at some early time some priest or priests organized the earliest version of the Roman religious calendar, which would have included specific days sacred to particular divinities. Although we cannot place credence in the later Roman tradition that assigned the origin of certain festivals to the various kings (e.g., the Parilia and Consualia to Romulus or the Terminalia and Fordicidia to Numa), some of the large-lettered festivals in the calendar are likely to derive from cults established under the kings and deriving from the earliest version of the calendar, into which they were incorporated according to the temple's day of dedication, just as happened in subsequent Republican times. Yet, despite our doubts regarding the historicity of later traditions concerning the kings' creation of cults, Diana's temple on the Aventine and August 13 as its day of dedication can probably be accepted as belonging to the second half of the sixth century B.C. and preceding the establishment of the Republic.

A second, corroborating pattern of early Roman calendrical conscientiousness is exhibited by the *Fasti Triumphales*, which always include the day on which the victorious commander celebrated his triumph. Although this list was not inscribed until the time of Augustus, its data must have ultimately derived from archival material; and once again, although we should not accept all of its information dating before c.300 B.C. as entirely accurate, we likewise should not dismiss it all as completely fabricated in later times.

Parallel to the tradition of recording the dates of triumphs is the Roman habit of noting the days upon which the Roman state enjoyed smashing victories or suffered catastrophic defeats in war. Unfortunately, our surviving ancient sources mention calendrical dates for important battles very sporadically. Nevertheless, what we do have suggest strongly that recording and remembering these dates was a standard feature of Roman officialdom. The earliest and most famous example of such a battle day is, of course, July 18, the Day of the Allia (*dies Alliensis*). On that day in 390 B.C., the Romans were defeated by the Gauls, who then followed up their victory by occupying the city of Rome itself. The day was still being marked 400 years later in the epigraphic calendars of the early principate.[1] It is, however, generally overlooked by modern scholars that we possess the record of an even earlier battle day, which further testifies to early Roman interest in recording such matters. Ovid (*Fasti* VI.721–724) writes that June 18 was the anniversary of Tubertus inflicting a defeat upon the combined forces of the Aequi and Volsci. In order to understand Ovid's brief allusion, we must turn to Livy (IV.26–29), where under the year 431 B.C. he devotes considerable space to describing how A. Postumius Tubertus scored a signal victory over the Aequi and Volsci, who had raised forces under a *lex sacrata*. Livy's account is further embroidered by a tale involving the dictator executing his own son for disobeying orders. As Livy himself recognizes, the story is patterned after the famous one of Manlius Torquatus and his son in 340 B.C. (see Livy VIII.7). Although Livy's narrative for this episode in 431 B.C. has suffered considerably at the hands of his annalistic predecessors, who have created a largely fictitious account, we need not doubt that the brief reference in Ovid derives from Roman antiquarian literature and reflects the official memory of an actual event in the far distant past.

ROMAN CLASSIFICATION OF THE DAYS

Our most detailed and systematic treatment of the ancient Roman classification of the days of the year is Macrobius (*Saturnalia* I.16.2–35). Not all the details recorded in this passage are relevant to the present study, but those that are can be summarized as follows. A day was either a *dies festus* or a *dies profestus*. The former were festivals on which there occurred sacrifices, banquets, or games in honor of the gods. The latter were days available for humans to carry out necessary work, chores, and other business, both public and private. These days were also further subdivided into two groups: *dies comitiales* and *dies fasti*. The former were days on which it was lawful for the Romans to be summoned to an assembly in order to vote on legislative proposals or to elect magistrates. The latter were what we might term court days, because on them an elected official, the praetor, who from 366 B.C. onwards supervised litigation in the Roman community, made himself available to the public to grant the right for parties to litigate (with the appointment of a third person to serve as judge of the dispute) and to pronounce judgments in other matters. Roman antiquarians explained the term *fastus* as deriving from *fari* (to speak) and meaning that on *dies fasti* the praetor was allowed to utter the three legally important words *do* (I grant), *dico* (I decree), and *addico* (I assign).[2] In addition, if there were no assembly of the people, the praetor could conduct legal business on *dies comitiales*, so that the latter were in a sense also *dies fasti*; but the converse was not true: public assemblies could not be convened on *dies fasti*. Moreover, certain festivals (*dies festi*) were classified as *nefasti*, days on which the praetor was not allowed to perform any legal activities that required him to utter the three words mentioned above; nor could assemblies be held on these days. After observing that *dies nefasti* were not to be polluted by the performance of ordinary work, Macrobius proceeds to give varying expert opinions that qualified and relaxed this strict interpretation in different ways. He further points out that this stricture had to be rigidly observed by the *rex sacrorum* and the *flamines*, and that in order to prevent these priests from seeing work being done on holidays, their progress in public was preceded by a crier whose duty was to announce the coming of the priest, so that people engaged in work could cease until he had passed by.[3] Finally, Macrobius mentions another category of days, the *dies intercisi* (the split days), which were partly *fasti* and partly *nefasti*. On them public sacrifices were performed, but the day was *nefastus* in the morning and in the evening as the victim was slain and then later had its entrails formally offered up, but in the interval the day was *fastus*.

Besides marking the dividing days and listing the nundinal days and the abbreviated names of festivals and divinities, the surviving inscribed and painted Roman calendars use another series of letters to classify the character of each day.[4] The letters C, F, N, and EN were used respectively to characterize a day as *comitialis*, *fastus*, *nefastus*, or *endotercisus* (an archaic spelling of *intercisus*). These four abbreviations were used in Republican times to characterize the days in the calendar, but the epigraphic texts of the calendar also employ a fifth abbreviation: NP. Since we possess no explanation for this designation in extant ancient literature, its precise meaning is still a matter of modern scholarly debate. Although there is general agreement that the N must mean *nefastus*, there have been different proposals as to the signification of the letter P, and thus how this group of *dies nefasti* differed from the others.[5] In any case, of the 355 days of the pre-Julian calendar the character of 309 are certain, but plausible conjectures based upon patterns in the months can be used to postulate with reasonable confidence the character of the remaining 46 days. The result yields the following overall distribution: 195 C, 42 F, 58 N, 49 NP, 8 EN, and 3 *fisi*.[6]

It is important to remember that what we now possess concerning the character of the days of the year dates to the late Republic and early Empire. It probably has no relationship at all to the earliest Roman calendar of the regal period and perhaps not even of the entire early Republic. Michels (1967 52 and 106ff) has plausibly postulated that the original distinction involving the character of the days was simply between *dies fasti* and *dies nefasti*, and that *dies comitiales* were a later development. Indeed, the need for the latter during the regal period in the sixth century B.C. is likely to have been rather limited and could have been easily accommodated by a rather elastic notion as to the meaning of *dies fasti*. On the other hand, the preponderance of *dies comitiales* in the later fully developed calendar is certainly in keeping with the political needs of the Roman Republic during much of its history. Although only a portion of these days are likely ever to have been used throughout the course of a year, having a large number of days available for convening assemblies gave the Roman community considerable flexibility in meeting its political needs in the face of complicating factors such as weather, the demands of agriculture, and the military campaigning season. In addition, the large number of *dies comitiales* that were not used for assemblies supplemented the mere 42 *dies fasti* to afford Romans plenty of opportunity to take care of their legal business before the praetor.

The distribution of the *dies nefasti* is most interesting. Of the 58 total in the pre-Julian calendar, 45 occur in four long consecutive series: 13 in February interrupted only by the one NP day of the ides and followed immediately by another NP day (the Lupercalia); 14 in April interrupted only by the four NP days of the ides, the Fordicidia, the Cerialia, and the Parilia; ten in June leading up to QSDF on the 15th and interrupted only by the NP day on the ides; and eight in July interrupted only by the NP day of the Poplifugia. Moreover, since the NP days were themselves *dies nefasti*, to say that these four series were interrupted by such days is actually a misnomer. The series in early June offer a clue to understanding this peculiar distribution, as well as of the significance of the *dies nefasti* in the cultural environment of the late Republic and early Empire. As noted, the sequence of ten *dies nefasti* led up to June 15, the day on which the Vestals formally cleaned out the Atrium Vestae and disposed of the refuse (Ovid *Fasti* VI.713–714). Since the Vestals must have been quite busy throughout the first half of june in preparation for and then observing the Vestalia of the 9th, we may surmise that the series of *dies nefasti* during this period was designed to apply primarily to them as representatives of the Roman people as a whole, and who were required during these days to neither look upon nor to engage in ordinary work except as was necessary for the conduct of their obligatory religious rites. The same is likely to have applied to the series of *dies nefasti* in February and April. In the latter case, the series of days preceded and covered the period that encompassed the Fordicidia and the Parilia, which involved the participation of the *rex sacrorum* and the chief Vestal. The series in February begins with the first day of the month and leads up to the great purification of the Lupercalia, so that it seems to resemble the pattern in June with strictures on ordinary work being applied narrowly to various priests engaged in the necessary preparatory rites. Although this interpretation of the *dies nefasti* may not successfully explain all such days,[7] it can account for the great majority and may also provide us with another datum concerning the evolution of the Roman calendar as a reflection of the practical needs of the Romans: for applying the strictures of *dies nefasti* rather narrowly to a small group of religious experts would have served the needs of a large and complex community in which ordinary people needed to be allowed to go about their regular business with a minimum of interruption, while priests and magistrates performed the necessary religious functions on behalf of the entire community.

This point leads us to the question of the 49 NP days. Their distribution is as follows: The ides of all 12 months are thus labeled, as is March 1, originally the first day of the Roman year. The remaining 36 days marked NP belong to that very special class of 48 days whose abbreviations are marked in the epigraphic calendars with large letters. It therefore seems rather obvious that NP designated *dies nefasti* that were somehow more special than the others. As Wissowa (1912 438–439) and Michels (1967 76–77) have argued, the P in NP may stand for *publicus* and indicates that the day was to be a general holiday on which the public at large was expected to refrain from its usual labors to rest and, if so inclined, to participate in the religious activities of the day.

Finally, the eight *dies intercisi* form a curious pattern in the calendar as well. The two Equirria are preceded by EN days, whereas the remaining six come between days that are NP in character. Seven of the *dies intercisi* are even numbered days, and the only one that is an odd numbered day is the one that precedes the Equirria of March 14, the sole large-lettered festival that falls on an even day.[8] It is difficult to say exactly what this pattern means. Since the *dies intercisi* were days on which sacrifices were formally offered, perhaps they served as *feriae succidaneae* or *feriae praecidaneae* for the NP festivals that they followed or preceded. Gellius (IV.6.10) cites the fifth book of Ateius Capito's work on pontifical law concerning Ti. Coruncanius, who was *pontifex maximus* around the middle of the third century B.C., as having held a *feriae praecidaneae* on a *dies ater*, which required the pontifical college to rule that doing so (i.e., holding it on a *dies ater*) would not be regarded as taboo (*religioni*).

THE AFTER DAYS[9]

One curious group of days in the Roman calendar was that of the after days, *dies postriduani*. They were 36 days, each one being the day immediately following one of the three dividing days of the 12 months. The adjective *postriduanus* derived from the adverb *postridie*, meaning 'on the day after' and hence the author's translation of *dies postriduani* as the after days. By the time that the Romans began developing their own native literature, these days were also termed *dies atri*, the black days, and were considered ill-omened. Excluding Gellius (IV.6.10) mentioned at the end of the preceding paragraph and pertaining to the middle of the third century B.C., our earliest reference to a *dies ater* dates to the second year of the Hannibalic War (217 B.C.) after the Roman disastrous defeat at Lake Trasimenus and occurs in a passage of Livy (XXII.10), in which he records a lengthy formula used by the *pontifex maximus* in obtaining the Roman people's permission to vow a sacred spring in order to restore the *pax deorum* (goodwill of the gods) and thereby to achieve victory over the Carthaginians. The pontiff's formal vow contained various exceptions, so that the extraordinary offering to the gods would not be vitiated through technical faults. One of these conditions reads as follows: "if he [any Roman] shall have unknowingly offered it on a black day, it shall be done properly = si atro die faxit insciens, probe factum esto." Our next datable encounter with this group of days occurs in a passage of Macrobius (*Sat.* I.16.21–24), in which he gives an explanation for the origin of the ill-omened nature of these days. In doing so, he cites two Roman historians: Cassius Hemina and Cn. Gellius. The former probably wrote his work during the third quarter of the second century B.C., whereas the latter seems to have composed his historical account during the decade or two leading up to the Social War (90–88 B.C.). The passage reads as follows:

> Gellius in the fifteenth book of his annals and Cassius Hemina in the second book of his history record the reason for these [sc. *dies postriduani*]. 22. In the 363rd year from the founding of the city the military tribunes Verginius, Manlius, Aemilius, Postumius, and their colleagues had it discussed in the senate why it was that the state had been sorely afflicted so many times within a few years. 23. L. Aquinius, a *haruspex*, having been ordered by decree of the fathers to come into the senate for inquiries into religious matters, said that the military tribune Q. Sulpicius, when about to fight against the Gauls at the Allia, performed a sacrifice on the day after the ides of Quinctilis [July 16] for the sake of the fighting; and that also at the Cremera and at many other times and places, following a sacrifice on an after day, an engagement had turned out badly. 24. The senate then ordered that this matter concerning religion be referred to the college of pontiffs; and the pontiffs ordained that all days after the kalends, nones, and ides be regarded as black days, so that they might not be battle days, pure days, or days for assembly.[10]

We find the same explanation given for the nature of the *dies postriduani* in V.17 of Aulus Gellius' *Noctes Atticae*. Instead of mentioning the *haruspex* L. Aquinius, Gellius gives the man's name as L. Atilius and does not assign him any official status, but his statement concerning the military tribune Q. Sulpicius so closely resembles the wording of Macrobius that the two writers have obviously drawn their information from the same source; and thanks to Gellius citing his authority, we know it to have been the fourth book of Verrius Flaccus' *De Verborum Significatu*. Livy tells the same story of the after days in his own way at the beginning of his second pentad (VI.1.12), where under the year 389 B.C. he describes various measures taken by the Romans in the immediate aftermath of the Gallic capture of the city during the previous year. In fact, what Macrobius records in the passage quoted above became the canonical explanation for these days. Since Macrobius cites Cassius Hemina and Cn. Gellius jointly for the entire story, we have no way of telling whether there were any significant differences between the two authors on this matter. On the other hand, the join citation could indicate that Cn. Gellius may have simply borrowed the tale from Hemina without introducing major changes. We may therefore assume that this explanation for the *dies postriduani* was current at least as early as the third quarter of the second century B.C. in the historical work of Cassius Hemina.

There is, of course, a major problem with this explanation for the after days. The battle of the Allia, which was generally supposed as having been responsible for the taboo on all 36 after days, was not fought on July 16, the day following the ides, but on July 18. Thus, in order to maintain the logic of this explanation, ancient writers generated two slightly different versions of the events of 390 B.C. One version (e.g., that of Macrobius, Gellius, and Livy just cited) accounted for the two-day discrepancy between July 16 and the battle day of the Allia by portraying the Roman commander in the field as having performed the pre-battle sacrifice on a *dies postriduanus* two days in advance. The other version asserted that the battle was actually fought on July 16, and that the Gauls had occupied Rome itself two days later on July 18, thus accounting for both the 36 after days and the damning of July 18 in later times as an ill-omened day associated with the year 390 B.C.[11] Since Polybius (II.18.2) says that the Gauls occupied the city two days after the battle, it seems likely that this second version of events was current when Polybius was writing, probably about the same time as Cassius Hemina. In any case, neither version is convincing. It is very unlikely that the Roman commander would have conducted a sacrifice to obtain favorable omens two days in advance and then not again on the actual day of the engagement. The alternative explanation is also flawed in that the epigraphic calendars along with Livy (VI.1.11) and Tacitus (*Hist.* II.91) agree in marking July 18 as the anniversary of the battle at the Allia, not the capture of the city.[12] There is another major logical objection to the later ancient explanation for this group of 36 days. Why did the pontiffs not rule that July 18 alone was to be regarded henceforth as an ill-omened day? Why did they include in their ruling this entire group of 36 days? The ancient accounts were clearly aware of this objection and tried to explain it away by claiming that there had been other disasters to the Roman state that had caused the pontiffs to classify all

36 after days as ill-omened. The implication, of course, is that these other disasters occurred on other *dies postriduani*. In fact, in order to explain the pontifical ban on pre-battle sacrifices on an after day in the future, Gellius (V.17.2) portrays some senators in 389 B.C. as remembering that other pre-battle sacrifices had been conducted on after days and were followed by Roman defeats. In sum, the ancient canonical accounting for these days is obviously inadequate, and we should search for an alternative explanation.

Two similar passages from Roman antiquarians, who were keenly interested in the history of the meaning of words, provide us with a far more satisfying explanation for the after days. In his all too brief catalogue of the derivations of the names of Roman festivals Varro writes as follows concerning the Quinquatrus of March 19: "Quinquatrus: as the result of a mistake with its name, this single day is observed as if it were said to be five (*quinque*). Just as the Tusculans similarly call the sixth day after the ides *sexatrus* and the seventh day after *septimatrus*, so this one, because it was the fifth day after the ides, is *quinquatrus*."[13] A similar explanation for Quinquatrus is given by Festus, whose information derives from Rome's other great antiquarian, Verrius Flacus of the Augustan age: "the shape of that word is uttered by the example of many Italian peoples, because that festival is the fifth day after the ides, just as among the Tusculans there are *triatrus, sexatrus*, and *septematrus*, and among the Faliscans *decimatrus*."[14] Both Varro and Verrius Flaccus were combatting a later Roman misinterpretation of the Quinquatrus, which regarded it as a festive period of four days, March 16–19 and coming immediately after the ides. In fact, by the late Republic the Quinquatrus was no longer a single day (March 19) in honor of Minerva, as it had originally been, but it had become a four-day holiday (Ovid *Fasti* III.809–814). The two quotations from Varro and Verrius Flaccus make it clear that Quinquatrus was a compound word formed from *quinque* and *atrus*, the latter being a dialectal variant of *alter*, meaning 'other' or 'next'.[15] Thus, Quinquatrus originally meant "Fifth-Other" or "Fifth-Next," just as *triatrus, sexatrus, septimatrus*, and *decimatrus* meant respectively "third-next," "sixth-next," "seventh-next," and "tenth-next." It therefore seems quite obvious that *dies ater*, a synonym for *dies postriduanus*, originally stood for *dies alter* and simply meant the next or following day, hence its equivalence to *dies postriduanus*. The term also would have been applicable to all 36 dividing days of the year and thus would have easily formed their own category of 36 *dies postriduani* or *dies atri*. At some time, however, most Romans lost sight of the original meaning of dies atri and reinterpreted them to mean "black days" and hence ill-omened days. Eventually the battle of the Allia and the capture of Rome by the Gauls was used to provide these days with a suitable, but erroneous, explanation.

Having dealt with these days in terms of Latin linguistics, let us turn to the other obvious question regarding these days. Why were they considered ill-omened? Immediately following the passage quoted above, Macrobius adds: "but also, Fabius Maximus Servilianus, the pontiff, in his twelfth book, says that one should not sacrifice to ancestral spirits on a black day, because then too it is necessary to invoke first Janus and Jupiter, who must not be named on a black day."[16] The Servilianus cited here was consul in 142 B.C.; and besides having written a history of Rome, he seems to have composed a work on pontifical law, from which this statement has been taken.[17] Since Servilianus was a contemporary of Cassius Hemina, it would be interesting to know how his injunction against performing a *parentatio* on a *dies ater* related to Hemina's explanation of the 36 after days. Unfortunately, Macrobius' brevity does not allow us to settle this matter. Servilianus might have accepted the explanation as we have it in Hemina, and he could have simply been providing additional information concerning these days in terms of actual cultic practices. On the other hand, his position as pontiff may have given him an entirely different perspective on the matter and could have convinced him that the after days had nothing at all to do with the events of 390 B.C. In any case, Servilianus' remark fortunately provides us with the necessary clue for understanding why the after days were regarded as ill-omened, irrespective of their later misinterpretation as black days.

The key here is Servilianus informing us of the general rule that Janus and Jupiter could not be invoked on a *dies ater.* To the Romans, these two divinities, more than any others, embodied primacy. Like the Greek Zeus, Jupiter, as lord of the sky and the weather, was recognized as the king of the gods; and his preeminent position was shared in the Roman state religion by his priest, the *flamen Dialis*, who was ranked first among all the fifteen *flamines*. Moreover, as god of the sky and light, the ides of every month were sacred to him, because when the Romans were using a lunar calendar in very early times, the ides corresponded to the time of the full moon when the night sky was most illuminated (Macrob. *Sat.* I.15.15–18). Janus, besides being a god of doorways, was regarded in more abstract terms as presiding over beginnings of all sorts. Hence, the kalends of every month were sacred to him (Macrob. *Sat.* I.9.16). In addition, he was viewed as the divinity who granted mortals access to the other gods and hence had to be invoked first (Ovid *Fasti* I.171–174 and Macrob. *Sat.* I.9.3). Janus' control over beginnings was even enshrined in Roman myth that portrayed him as the first being who ruled over Latium.[18] The primacy that Janus and Jupiter enjoyed in Roman religion is illustrated by the prayer that Cato (*De Agri Cult.* 141) records for conducting a purificatory *suovetaurilia* of one's farmland. Although Mars was the divinity invoked for the ritual, the prayer begins with an invocation to Janus and Jupiter. We encounter the same phenomenon in Livy's quotation of the *devotio* formula (VIII.9.6), in which P. Decius Mus first calls upon Janus and then the three gods of the major flaminates (Jupiter, Mars, and Quirinus) before invoking the other divinities. Given the centrality of the three dividing days in defining the structure of each month, and from which the Romans reckoned all calendrical dates, the dividing days enjoyed the same primacy in the temporal sphere as did Janus and Jupiter in the divine one. Consequently, associating the two would have been quite natural. On the other hand, the 36 after days, all coming after a dividing day and falling upon an even numbered day, must have been regarded by early Romans as embodying a quality opposite to the primacy of the dividing days and of Janus and Jupiter. Their "afterness" would have been viewed as incompatible with the "firstness" of the two gods, hence the rule mentioned by Servilianus that these two divinities must not be invoked on the after days.[19]

We now finally turn to the matter of trying to determine exactly what the nature of the dies postriduani was. As the embodiment of afterness and the antithesis of Janus' and Jupiter's firstness, we may postulate that they were most logically associated with gods of the netherworld; and this supposed connection leads us into another Roman legal and religious classification of things. By the late Republic, Roman private law had become a well-developed discipline, subdivided into discrete areas, each containing further subdivisions and numerous formal definitions. One such major area of the private law was property, the law of things (*res*). In introducing his readers to this subject Gaius (*Inst.* II.2–9) briefly sets forth a series of standard juristic definitions of divine things before engaging in his detailed legal analysis of property owned by people. It can be summarized as follows. Things belong either to the gods (*res divinae*) or to people (*res humanae*). The former consists of three subdivisions: things that belong to the gods above (*res sacrae*), things that belong to the gods below (*res religiosae*), and things that are placed under the protection of the gods (*res sanctae*).[20] The only major thing that Roman jurists regularly assigned to this last category were the walls of towns, akin to the sanctity of the Roman *pomerium*. *Res sacrae* comprised temples and numerous other things that were dedicated or given to the gods. *Res religiosae* largely consisted of tombs; and a *locus religiosus* could be created simply by placing the remains of a dead person in the ground or in a structure above ground. A passage in Festus, however, further demonstrates that days of the year, as well as physical things, could be *religiosi*:

> *religiosus* is not only considering important the sanctity of the gods, but also being dutiful toward humans. Moreover, dies religiosi are those on which it is regarded as forbidden to do except what is necessary: days such as the 36 called *atri*, the *Alliensis*, and those on which the *Mundus* is open. Aelius Gallus ** to be **, because it is not permitted for a person to act in such a way that if he does it, he may seem to be acting against the will of the gods. In this category are these things: for a man to enter the temple of Bona Dea; ** against ** mystic **; to propose a law to the people; to litigate before the praetor on a *dies nefastus*. Moreover, he very elegantly records the differences among *sacrum, sanctum*, and *religiosum*. He says that it is agreed that *Sacrum* is a building consecrated to a god; that *sanctum* is the wall that is around a town; and that *religiosum* is a tomb in which a dead person has been entombed or buried.[21]

The Aelius Gallus cited here was C. Aelius Gallus, a scholar of the late Republic or early principate, who was learned in the law. Gellius (XVI.5.3) cites the second book of his treatise "concerning the meaning of words that relate to the civil law = de significatione verborum quae ad ius civile pertinent." he is cited elsewhere in Festus. As can be seen from the asterisks, the text of this passage is lacunose; and the remaining few lines of the entry, whose meaning cannot be easily deciphered, have been omitted without affecting the conclusions reached here. Despite the gaps in the quoted text, its overall gist is not hard to follow. After first giving a general definition of *religiosus*, the text applies the concept to three distinct things: days that are *religiosi*, human acts that are *religiosi*, and places that are *religiosi*. With respect to the latter two categories the passage cites Aelius Gallus, whose views continue to be set forth in the remainder of the quoted text concerning the three-fold distinction of *res sacrae*, *res sanctae*, and *res religiosae* that clearly derive from Roman juristic literature. The passage is important in confirming that the *dies postriduani* were classified by Roman experts in the law and religion as *religiosi* and hence associated with the gods of the underworld.[22] The connection helps to explain further why the Romans termed the after days *atri* and considered them ill-omened. As we have seen, the designation of these days as *atri* most likely began as the result of using a dialectal variant for *alter*; and it just so happened that the variant form was a homonym for the word black. Was the verbal change the result of the Romans associating the color black with *res religiosae*? Or did the verbal change come first, and the coincidence of *ater* meaning both 'next' and 'black' simply reinforce an already existing notion that the after days were ill-omened? However the terminology of *dies atri* and their meaning evolved, there can be no doubt that the blackness of these days contributed in later times to their perceived character as *religiosi*.

The following five passages are the most important ancient statements concerning what could or could not be done on *dies atri* and *dies religiosi*; and from them we can gain a clearer understanding of their nature.

1. Varro, "The days following the kalends, nones, and ides were called black, because during those days people were to begin nothing new."[23]
2. Macrobius, "Moreover, according to our ancestors, the after days, which they also damned as black as if with an unlucky name, had to be avoided for everything."[24]
3. Gellius, "The Pontiffs [389 B.C.] decreed that no sacrifice would be proper on these days."[25]
4. Gellius, "*Dies religiosi* are so called for being infamous and impeded by a gloomy omen, on which one must refrain from performing religious rites and embarking upon anything new."[26]
5. Macrobius, "Hence Varro too writes thus: 'when the *Mundus* is open, it is as if the doorway of the somber and nether gods is open. Wherefore, it is taboo (*religiosum*) not only to engage in battle, but also to hold a military levy and for a soldier to set out, to send off a ship, and to take a wife for the sake of having children.'"[27]

Texts 1 and 2 agree in asserting that on after days there was a general ban on new undertakings and performing religious rites. Text 3 confines itself to mentioning a ban on conducting sacrifices. Text 4 is speaking of *dies religiosi* in general instead of *dies atri*, but what it says is consistent with the preceding three texts in stating both a general ban on new undertakings and on religious activities. Text 5 concerns itself with a particular set of three *dies religiosi*, the days on which the *Mundus* was opened, namely, August 24, October 5, and November 8 (see Festus 144–145L s.v. *Mundus* with Fowler 1912). Nevertheless, its overall tenor is consistent with what else we know about the *dies religiosi* and the subgroup of *dies atri*; and it is informative in that instead of simply stating a general ban, it gives examples of the sorts of undertakings that should not proceed on these days. Within this short list are included a ban on engaging in battle, levying an army, and entering a marriage for starting a family,[28] but these activities are associated in Text 5 not with the *Dies Alliensis* and the *dies postriduani*, but with the days on which the Mundus was open. Since all these bans are likely to have applied to all *dies religiosi*, the later Roman explanation of the *dies postriduani* in terms of the events of 390 B.C. is clearly seen to be an erroneous fiction. Finally, it should be noted that Michels (1967 66) observed that there is no evidence for a Roman triumph during Republican times being celebrated on a black day.

In conclusion, the after days and their attendant taboos were of great antiquity and were the product of Roman attitudes toward the nature of time, particularly as it was structured in their calendar. Their origin had nothing to do with the events of 390 B.C., but the close proximity of the Day of the Allia, July 18, to the after day July 16 prompted later Romans to use the former in order to explain the entire set of 36 after days. We need not doubt that the Gallic defeat of the Romans at the Allia and the subsequent occupation of Rome itself had a tremendous impact upon the Romans, both then and henceforth; and that in the aftermath of the disaster Roman priests did their share in the reconstruction by not only reestablishing the *pax deorum*, but they also must have examined recent events in an attempt to determine what had gone wrong and to take measures to ensure that the catastrophe would not be repeated. Placing the *Dies Alliensis* among the *dies religiosi* was certainly one of these measures. Another one involved the censorship. Under the year 392 B.C. Livy (V.31.6) writes: "the censor C. Julius died, and into his place M. Cornelius was made suffect. Afterwards this matter was taboo (*religioni*), because Rome was captured in that *lustrum*, and henceforth a censor is never substituted into the placed of a deceased one."

EXCURSUS ON THE DAY OF THE CREMERA

Before moving on to consider the interesting history of the black days during the late Republic, it is necessary first to pause to explore another related matter, the Day of the Cremera (*dies Cremerensis*). By the time that Livy and Dionysius of Halicarnassus came to write their early histories of Rome at the beginning of the Augustan age, there already existed in the annalistic tradition the fully developed story of how in 478 B.C. 306 members of the Fabian family had shouldered the entire burden of Rome's war against Veii only to be annihilated in a battle at the Cremera and leaving behind a single male to carry on the Fabian name.[29] Relevant to this study is one later Roman tradition that maintained that the battles at the Allia and the Cremera had occurred on the same day (Livy VI.1.11, Plut. *Cam.* 19.1, Tac. *Hist.* II.91, and Macrob. *Sat.* I.16.23). Besides these literary *testimonia*, one of the epigraphic calendars, the *Fasti Antiates Ministrorum Domus Augustae* (Degrassi 1963 208), dating to the reign of the Emperor Tiberius, labels July 18 as "the Day of the Allia and of the Fabii = Dies Allia[e et] Fab(iorum)". On the other hand, there was an alternative tradition according to which these two disasters were linked in the calendar in a different way. Rather than the two battles sharing the same anniversary, the destruction of the Fabii occurred on February 13, the same day on which the Gauls ended their occupation of Rome. The evidence for this alternative synchronism is to be assembled from the following three data.

1. Plutarch (*Cam.* 30.1) says that the Gauls entered the city shortly after the ides of July, occupied it for about seven months, and abandoned it on the ides of February.
2. According to Ovid (*Fasti* II.195–196) the Fabii fell at the Cremera on the ides of February.
3. The calendar of Polemius Silvius of 449 A.D. (Degrassi 1963 265) characterizes February 13 in these words: "the *parentatio* of graves begins, on which day Rome was freed from the siege of the Gauls = parentatio tumulorum inc(ipit), quo die Roma liberata est de obsidione Gallorum."

Two important conclusions can be reached from these data: (I) given the fact that the sources disagreed as to when the battle at the Cremera occurred, no one in later historical times had any authentic information upon which they were basing their dating; and (II) later Roman writers were in the habit of linking these two disasters, probably in order to flesh out their accounts of the Cremera legend with the slightly more authentic traditions surrounding 390 B.C. The latter point is well illustrated by a passage from Livy (IX.38.15–16) for the dictator year 309 B.C.: "Papirius declared C. Junius Bubulcus to be his master of the horse. While he was seeing to the passage of the *lex curiata de imperio*, a grim omen cut short the day, because the first to vote was the curia Faucia, distinguished by two disasters, the capture of the city and the Caudine Peace [321 B.C.]: for in both years the first vote had been that of the same curia. 16. Licinius Macer renders the curia abominable by even a third disaster, the one received at the Cremera."

Since ancient Roman historians, if left to their own devices, were inclined to invent Roman victories and not defeats, it is possible that at the heart of this legend lies the very faint memory of a Roman reversal at the Cremera, but the story, as we now have it, can hardly be regarded as historical. If Taylor (1960 40–41) was correct in placing the *tribus Fabia* upstream on the right bank of the Tiber near the Cremera, we can understand how historians decided to make members of the Fabian family the heroic martyrs of their story. Moreover, if there was no clear evidence as to when this disaster was supposed to have occurred, this too could have been suggested to Roman historians by the extraordinary series of seven consecutive Fabian consulships during the years 485–479 B.C.; and since this in turn coincided with Xerxes' invasion of Greece, the heroic deaths of the 300 Spartans at Thermopylae provided Roman historians with the necessary framework in which to construct their own tale of Roman heroism.

Associating the Cremera and the Allia is likely to have resulted from two things. One would have been the obvious fact that they were the two most significant military disasters of the early Republic; and since the calendrical date for the Allia alone was known, it was also assigned to the Cremera. The other cause for the connection between the two events might lie in another curious aspect of the consular *fasti*. During the period 405–367 B.C. the Romans every year (with only a few exceptions) elected six military tribunes with consular power to head the state; and although there are some instances in which two members of the same family held office together, there is only one instance in which three members of the same family did so; and this occurred in 390 B.C. and was done by the Fabii. It is therefore possible that some later Roman historians were encouraged to conflate the Cremera and Allia disasters because of Fabian preeminence in the consular *fasti* at the time of the two events. Evidence for historiographical conflation between these two episodes is offered by the role of the Porta Carmentalis in each tale. It was a gate in the so-called Servian Wall at the foot of the Capitoline Hill and took its name from the fact that it was situated near a shrine of Carmenta. According to Livy (V.46.9) the Gauls attempted their unsuccessful nocturnal ascent of the Capitoline at this point. Ancient sources also link the gate to the Cremera legend by explaining that the right-hand side of this gate was called *Scelerata*, because the Fabii had marched through it on their ill-fated way to the Cremera.[30] Lastly, the alternative dating of the Cremera to February 13 could have resulted from later antiquarians seeking historical aetiologies for the *Dies Parentales* of February 13–21 and of the Lupercalia of February 15, whose rites were carried out by two groups, the *Luperci Fabiani* and the *Luperci Quinctiales*. If so, Roman historians and antiquarians, who saw parallels between the Cremera and Gallic disasters, might have used this alternative dating of the former to arrive at the supposed day on which the Gauls departed from Rome.

In conclusion, it is quite apparent that little, if any, of the Cremera legend as related by the extant ancient sources can be given credence; and this certainly applies to later ancient claims as to the battle's calendrical date. Unlike the Day of the Allia and the anniversary of Postumius Tubertus' victory on

June 18 of 431 B.C. mentioned at the beginning of this chapter, the Day of the Cremera was obviously not known in later times.

BLACK DAYS AND BATTLE DAYS

The surviving ancient sources provide us with relatively few calendrical dates for battles fought by the Romans during the Republic preceding the outbreak of the civil war between Caesar and Pompey, but the few dates that we do have intersect in interesting ways with Roman views toward the *dies atri*. The sixth book of Ovid's *Fasti* differs from the other five in that it alone contains brief references to the dates of several battles, eight in all. It would be interesting to know what his immediate source for such information was. If we possessed his books for the months of July, August, and September, we would probably have many more such battle days. In addition to Ovid's eight dates, we have another ten recorded by Claudius Quadrigarius, Cicero, Livy, Velleius Paterculus, Plutarch, Appian, and Eutropius. All eighteen are listed below in chronological order; and following the date and brief description of the battle, a parenthetical reference first includes the source for the calendrical date, after which is given a citation of the most important ancient description of the battle itself. All dates, of course, are B.C.[31]

431: June 18, victory of the dictator A. Postumius Tubertus over the combined forces of the Aequi and Volsci (Ovid *Fasti* VI.721–724; cf. Livy IV.26–29).

241: March 10, Roman defeat of the Carthaginians at the Aegates Islands off the western coast of Sicily, ending the First Punic War (Eutrop. II.27.2; cf. Polyb. I.60–61).

217: June 22, defeat of the consul C. Flaminius by Hannibal at Lake Trasimenus (Ovid *Fasti* VI.763–66; cf. Livy XXII.4–7, and Polyb. III.80–85).

216: August 2, defeat of the Romans by Hannibal at Canae (Claudius Quadrigarius cited in Gell. V.17.4 and Macrob. *Sat.* I.16.26; cf. Livy XXII.41–49 and Polyb. II.110–108).

207: June 23, Roman victory over and death of Hasdrubal at the Metaurus River (Ovid *Fasti* VI.767–769; cf. Livy XXVII.43–50).

203: June 23, defeat and capture of Syphax by Masinissa (Ovid *Fasti* VI.767–770; cf. Livy XXX.11–15, and Appian *Punica* 26–28).

202: December 17, Roman defeat of Vermina, the son of Syphax, marking the last military action in the Second Punic War (Livy XXX.36.8).

168: September 4, victory of Aemilius Paulus over King Perseus of Macedon at Pydna (Livy XLIV.37.8; cf. Livy XLIV.40–43 and Plut. *Aem. Paul.* 18–21).

153: August 23, defeat of the consul Q. Fulvius Nobilior on the Volcanalia by the Celtiberians in Nearer Spain (Appian *Iberica* 45).

137: June 9, victory of D. Junius Brutus over the Callaici in Farther Spain (Ovid *Fasti* VI.461–462; cf. Livy *Per.* 55, *Per. Oxy.* 55, and Appian *Iberica* 72).

105: October 6, defeat of the Romans by the Cimbri and Teutones at Arausio (Plut. *Lucullus* 27.7; cf. Livy *Per.* 67 and Dio XXVII. fr. 91.1–4).

101: July 30, Roman victory over the Cimbri at Vercellae (Plut. *Marius* 26.4–5; cf. Plut. *Marius* 25–27).

90: June 11, defeat and death of the consul P. Rutilius Lupus by the Marsi at the River Tolenus (Ovid *Fasti* VI.563–566; cf. Apian *Bell. Civ.* I.43).

89: June 11, defeat and death of T. Didius by the Marsi (Ovid *Fasti* VI.567–568; cf. Appian *Bell. Civ.* I.40 and Vell. Pat. II.16.2).

82: November 1, Sulla's victory over his Roman and Samnite adversaries at the Colline Gate in Rome (Vell. Pat. II.27.1; cf. Plut. *Sulla* 29 and Appian *Bell. Civ.* I.93).

69: October 6, Lucullus' victory over King Tigranes of Armenia at Tigranocerta (Plut. *Lucullus* 27.7; cf. Plut. *Lucullus* 26–28).

53: June 9, defeat of M. Crassus by the Parthians at Carrhae (Ovid *Fasti* VI.463–468; cf. Dio XL.25–27, and Plut. *Crassus* 22–31).

51: October 13, Cicero's defeat of Cilicians or Syrians on Mount Amanus (Cic. *ad Att.* V.20.3–6).

Seven of the eight Ovidian battle days are clustered into three groups. The only one not paired up with another anniversary is the victory of the dictator A. Postumius Tubertus, the earliest battle day that we possess. The poet might have been attracted to it for inclusion in his poem because of its singular antiquity. Ovid balances the death and defeat of Flaminius on June 22 against the Roman victories over Hasdrubal and Syphax on the following day; and in doing so, he advises the emperor not to engage in battle on the former, but to wait until the latter. Thus, Ovid is likely to have chosen these days carefully in order to make this point. Similarly, his pairing of the two Roman defeats on the same day (June 11) in consecutive years (90–89 B.C.) clearly bears the message that superstition about the lucky or unlucky nature of days for conducting warfare was not the least bit frivolous; whereas his coupling of the victory of D. Junius Brutus and the catastrophic defeat of Crassus on the same day of the year is perhaps designed to convey a more complex and ambiguous idea: namely, that even on days favorable to the Roman state its commanders and soldiers must act responsibly and not trust in luck alone. Mention of Carrhae, of course, allows Ovid to praise Augustus for recovering the military standards lost by Crassus.

Three of these battle days come from Livy or the Livian tradition, and they are all alike in that they mark Rome's final victory in a major war: the battle of the Aegates Islands in 241, the last military engagement of the Second Punic War, and the battle of Pydna. This pattern clearly suggests that these felicitous events were recorded by the Roman state and its historians during Republican times. The second of these dates is of particular interest in this regard, because the final military action of the Second Punic War was a rather minor one, whereas the battle of Zama that had preceded it had been a large-scale battle fought between two of the greatest commanders of the ancient world; but unlike the minor engagement fought between Cn. Octavius and the Numidian Vermina, whose date Livy carefully records, we do not know the day on which Zama was fought.

A passage from Macrobius concerning the religious suitability of days on which to engage the enemy in battle offers us a conceptual framework in which to view these battle days. The text reads as follows:

> In levying men the ancients also avoided days that were marked by adverse things. They even avoided *feriae*, as Varro writes in these words in his books of the augurs: "One must not levy men on *feriae*. If he does, there is to be expiation." 20. Nevertheless, it must be understood that if the Romans were themselves bringing on the war, there was then the need for them to choose the day for fighting. But when they were the recipients of war, no day stood in the way of them defending either their own safety or public prestige: for what opportunity is there for observance when the ability of choosing is not present?[32]

Q. Claudius Quadrigarius, a Roman historian of the early first century B.C., is our only source for the date of Canae; and both Gellius and Macrobius cite him in exactly the same context, drawing their information from a common source, the fourth book of Verrius Flaccus' *De Verborum Significatu*. It is given in the form of a footnote to Gellius' and Macrobius' explanation for the origin of the *dies postriduani* in terms of the Day of the Allia. According to them, Quadrigarius asserted that as the result of Canae having occurred on August 2, four days before the nones, henceforth all fourth days preceding dividing days were regarded as unlucky and therefore constituted a second group of days parallel to the after days; but both Gellius and

Macrobius, doubtless repeating Verrius Flaccus, express doubt by observing that no such ban could be found anywhere in other writings. Their skepticism toward Quadrigarius' claim is well placed, because it is certainly his own fiction designed to sensationalize the Roman disaster at Cannae, similar to what we have seen in regard to Licinius Macer adding the Cremera disaster to those of the Allia and the Caudine Forks to the casting of the first vote by the curia Faucia in the curiate assembly in 309 B.C. What clearly escaped (or was simply ignored by) Quadrigarius is the fact that August 2 was in fact a *dies postriduanus*. If the dating of Canae to this day is correct, religious observance might have dictated that the Romans avoid engaging Hannibal on this day; but such a view neglects two important considerations. According to Macrobius, the Romans did not need to scruple about such matters if they were on the defensive. Did this apply to the war as a whole, or only to the immediate situation? If the former, the requirement of choosing a religiously suitable day for fighting would not have been applicable if the Romans regarded the enemy as responsible for the outbreak of hostilities; and this was certainly the case in the Roman mind with respect to the Hannibalic War. On the other hand, if the rule was applied more narrowly to the immediate circumstances preceding a battle, then a second possible consideration enters the equation: namely, the question of military exigency and the complexities of military operations resulting from the hurley-burley interaction with the enemy, who, of course, had no regard for the niceties of Roman priestly injunctions.[33] Indeed, Polybius' account of Canae, which is less melodramatic than Livy's, indicates that the battle came about as the result of complex interactions between the Romans and Carthaginians during the previous three days.

This brings us to the defeat of Q. Fulvius Nobilior in 153 B.C. Appian states that as the result of this serious reversal, Roman commanders henceforth would not engage the enemy on this day unless forced to do so. Two interesting points follow from this statement. First of all, it looks as if in the aftermath of this battle the Romans made an official decision concerning the ill-omened nature of August 23. Given what we know about how the state dealt with reports of prodigies and their expiation, it is likely that the issue was first raised in the senate and then handed over to religious experts (probably the pontiffs) to offer their advice, which the senate then accepted and cast into the form of a senatorial decree, just as the sources describe the senate behaving in the aftermath of the Gallic capture of the city in ruling upon the ill-omened nature of the 36 after days. Cassius Hemina would have been writing his history around this time, and the creation of this new *dies ater* (for that was certainly what it was termed) by decree of the senate may have prompted Hemina to account for the 36 after days in a similar manner. Secondly, Appian's description of the prohibition henceforth to engage the enemy in battle on this day looks as if the day was officially ruled *religiosus* with respect to military matters, but with the typically Roman sensible condition that soldiers could fight if the situation required. Finally, it is curious to note that the man, whose defeat occasioned this calendrical innovation, was the son of the M. Fulvius Nobilior (consul 189 B.C.), who had authored the first antiquarian treatise on the Roman calendar.

Plutarch's *Lucullus* 27.7 informs us that the battles of Arausio (105 B.C.) and Tigranocerta (69 B.C.) were fought on the same day, and the way in which Plutarch tells us of this coincidence is most interesting.[34] As Lucullus was laying siege to the Armenian city Tigranocerta, King Tigranes, who had not been inside the city, proceeded to raise a very large army of his allies. As he approached Tigranocerta to relieve it of the siege, Lucullus, whose forces were greatly outnumbered, was placed in a very difficult predicament. He decided to leave a skeleton force to guard the siege works while he engaged Tigranes in battle; but as he was about to carry out his plans, some advisors objected that he could not engage the enemy on this day, because it was the anniversary of Rome's crushing defeat 36 years earlier at Arausio. Plutarch explains that this day (October 6) was among the forbidden days called black (*melainas*), but Lucullus replied that he would convert the day into a lucky one (*eutyches*). He then proceeded to defeat Tigranes, whose vast army he virtually destroyed with very few Roman casualties, after which he captured Tigranocerta itself. He had succeeded in turning a situation of extreme military exigency into a stunning Roman victory. In calling this day 'forbidden' Plutarch employed the Greek adjective *apofrados*, derived from the preposition *apo* (= from) and the verb *frazein* (= to utter), so that the word corresponds to Latin *nefandus* (= unspeakable). It is therefore clear that Plutarch was attempting to translate Latin *nefastus* into Greek, but as Gellius (IV.9.5) observed, many Romans, who were unfamiliar with the precise meaning of words, often wrongly used *nefastus* to describe days that were properly termed *religiosi*. Plutarch was obviously guilty of this common error in this passage. Moreover, it emerges from Plutarch's words that following the disastrous defeat at Arausio, the day of the battle was officially ruled to be a *dies ater* and hence *religiosus*, doubtless with the same practical condition as Appian mentions in reference to Nobilior's defeat in 153 B.C.; and that Lucullus in face of a most dire military situation was therefore able to properly set aside the day's taboo nature. Unfortunately, Plutarch does not tell us whether as part of the honors granted to Lucullus by the senate for this dazzling victory, October 6 was declassified as a *dies ater*.

EPILOGUE

With the rise of autocracy under Julius Caesar's dictatorship and the subsequent establishment of the principate by Augustus Republican practices were forced to accommodate themselves to the new political reality. The Roman calendar was no exception. Two vestiges of imperial adulation are still with us today in the names of the months July and August, the original Quinctilis and Sextilis renamed to honor Julius Caesar and Augustus (Macrob. *Sat.*

I.12.34–35). Henceforth holidays were established to commemorate important days in the life of the imperial family (e.g., birthdays, accession days, and victory days). For example, an inscription from Cumae (*ILS* 108), dating to the reign of Augustus, and whose surviving text is not complete, enumerates no less than sixteen annual festivals to honor key events in Augustus' life. Yet, as emperors came and went over the course of the next few centuries, so did most of the days established in their honor. During the later years of Nero's reign the senate flattered the emperor by renaming April, May, and June respectively Neronius, Claudius, and Germanicus (Tac. *Ann.* XV.74 and XVI.12); and Domitian was similarly honored by having the months of September and October renamed Germanicus and Domitianus (Suet. *Dom.* 13.3 and Macrob. *Sat.* I.12.36; cf. Weinstock 1971 154–155). The surviving evidence of Imperial holidays is massive; and simply cataloguing it would require a large volume in its own right.[35] Nevertheless, the phenomenon is succinctly illustrated by three calendars, each separated from one another by about a century, which give us snapshots, as it were, of what Imperial holidays were currently important to the inhabitants of the Roman Empire: (1) the *Feriale Duranum*, a papyrus discovered at the Roman military town of Dura Europus near the Euphrates, dating to c.230 A.D. and listing festivals to be celebrated by the Roman soldiers; (2) the calendar of Furius Philocalus, dating to the year 354 A.D. when Roman paganism was rapidly giving way to Christianity; and (3) the calendar of Polemius Silvius, dating to 449 A.D. when the western half of the Roman Empire was disintegrating into a collection of much less civilized Germanic kingdoms.[36]

But as regards the overall content of this chapter, it needs to be noted that during the reign of Augustus the Day of the Allia was recycled and absorbed into the new political culture of imperial adulation. When Augustus' presumptive heir, his one surviving grandson and adopted son, C. Caesar, died of a wound on February 21 in 4 A.D. while campaigning against the Armenians, people throughout the empire were plunged into mourning. An inscription from Pisa in northern Italy (*ILS* 140) records a decree of the local senate, expressing the community's grief; and among the various yearly ceremonies established to commemorate the young man's death is the provision that henceforth February 21 be observed by the community as a sad day, like the *dies Alliensis*, with no public sacrifices, thanksgivings, marriage engagements, or public banquets.

3 The Rites of the *Argei*

A visitor in ancient Rome would have witnessed a curious rite performed on May 15. A group of pontiffs, Vestals, and magistrates progressed solemnly through the city to collect from specific sites thrice nine scarecrow-like figures called *argei*, fashioned out of rushes, after which the Vestals tossed them into the Tiber from Rome's oldest bridge, the *Pons Sublicius*, while a crowd of on-lookers may have exclaimed, "*sexagenarios de ponte* = sixty-year olders off the bridge!" The modern bibliography on this ceremony is extensive,[1] and interpretations have abounded since classical antiquity. The most common ancient explanation was that the thrice nine rush dummies were later substitutes for actual human victims sacrificed in a barbarous primitive past. According to one such view the transition from human sacrifice to the offering of lifeless anthropomorphic surrogates was first instituted by the Greek mythical hero Hercules during his supposed sojourn in Rome. Not only was this explanation consistent with the ancient notion of the civilizing effects of Hercules' travels among savage peoples, but it was also used to account for the curious designation of the rush puppets; for the latter, it was asserted, took their name from 'Argives', the appellation of Hercules' retinue (Greek *Argeioi* = Latin *Argivi*).[2] Festus, however, records another explanation whereby the rites of the *argei* were said to have been first observed at the time of the Gallic capture of Rome in 390 B.C. According to this alternative explanation a food shortage following the Gauls' departure from the city forced the Romans to reduce the population by throwing senior citizens into the Tiber, but one person, moved by filial piety, concealed his elderly father; and after it was learned that the latter had benefitted the state through advice in the person of his son, humanity finally prevailed when the Romans decided to offer rush puppets in place of aged citizens. According to this view the anthropomorphic figures took their name from the place where the son had removed (arcuisset < arceo) his father.

Despite the fanciful nature of these ancient etiologies, two of the earliest modern investigators into these rites patterned their own interpretation of the *argei* after the Roman notion that the rush dummies were substitutes for actual human victims. Wissowa argued more than a century ago that the rites were first introduced to Rome sometime during the First and Second Punic Wars when the Sibylline Books were consulted during a food shortage or pestilence. The Romans accordingly sacrificed 27 live Greeks by drowning them in the Tiber, but they subsequently maintained the observance of this extraordinary act by using figures made of rushes. Wissowa even followed the ancient tradition in deriving the term *argei* from Greek *Argeioi.*[3] Fowler (1902A and 1911 54 and 320–322), however, in the author's view irrefutably demolished Wissowa's thesis. First of all, he observed, if the ceremony had been introduced via the Sibylline Books, we would expect the *decemviri sacris faciundis* to have presided over them, but as the ancient *testimonia* make clear, the priests involved were the pontiffs, Vestals, and the *flaminica Dialis*, which strongly suggests great antiquity. Secondly, since our sources for this period record the Romans burying alive pairs of Greeks and Gauls (see Eckstein 1982), would they not have also reported this even more barbarous and sensational ceremony? Fowler's *argumentum e silentio* can be further strengthened by noting that Orosius, a Christian writer of the fifth century A.D. who epitomized and redacted Livy's history with the purpose of demonstrating the inhumanity of the pagan past, certainly would not have neglected to exploit such an act of savagery. Thirdly, it hardly seems likely that the poet Ennius, born in 239 B.C., would have attributed the rites of the *argei* to King Numa Pompilius (see Varro *L. L.* VII.43–44) if they had been of such a recent origin.

Clerici (1942) argued that the rites of the *argei* were first instituted in the aftermath of the Gallic capture of Rome. He derived *argei* from the toponym Arx (= the Citadel), the name which the Romans applied to the northeastern summit of the Capitoline Hill. Citing Livy's report in V.50.4 concerning the institution of the Capitoline Games in 390 B.C., Clerici conjectured that just as this celebration was undertaken to commemorate the Capitolium, the name for the southwestern summit of the Capitoline Hill, so the rites of the *argei* were established at this same time in honor of the Arx. Like Wissowa's reconstruction, Clerici's is open to various objections. First of all, it has already been demonstrated in the preceding chapter how the Day of the Allia was used by later Roman antiquarians to explain the origin of the 36 after days. In fact, other ancient authors assigned the institution of the Capitoline Games to Romulus.[4] Thus, it seems that these games were of such great antiquity that later writers really knew nothing about their origin. Secondly, Livy makes it clear that the *collegium Capitolinorum* charged with the duty of holding these games to honor Jupiter Optimus Maximus comprised the inhabitants of both the Capitolium and Arx.[5] Even if we assume that Livy's etiology for the games is unhistorical, his inclusion of both Capitoline summits in this celebration is likely to reflect the actual circumstances of later historical times. Thirdly, it is very unclear how the thrice nine *argei*, whose stations were situated throughout the city, were closely connected with the Arx. Finally, as will emerge below, a much better etymology for *argei* can be found than the one suggested by Clerici.

Since the later ancient explanations for the original religious significance of the *argei* appear to be little more than fanciful speculations, the temporal location of the rites of the *argei* in the official calendar probably offers us the best means by which we may come to understand their function in early Roman religion. Unfortunately, the ancient sources focus their attention almost entirely upon the disposal of the *argei* in mid-May, and this in turn has often resulted in modern scholars likewise concentrating almost exclusively on this one aspect of the *argei*; but it is important to realize that the ceremony of mid-May formed an epilogue, as it were, and not the central core of the rites. According to Ovid's *Fasti* III.791–792, a procession was made every year to the chapels of the *argei* on March 16 and 17. This came immediately after the festival of Anna Perenna of March 15, which celebrated the return of the new year, because, of course, in early times March 1 marked the beginning of the Roman year. During this month, the salii performed their dance throughout the city to drive out the old Mars (Mamurius Veturius) and to insure the growth of the crops through the sympathetic magic of their leaping. On March 1, Vesta's sacred fire was rekindled, and fresh laurel was used to festoon the temple of Vesta, the curiae, and the domiciles of the *rex sacrorum* and the *flamen Dialis* (Ovid *Fasti* III.135–166). These activities clearly demonstrate that the early Romans regarded March as a period of renewal and agricultural reawakening. Consequently, just as fresh laurel was used at this time for religious purposes, so it is likely that the procession to the chapels of the *argei* on March 16 and 17 involved the installation of new puppets fashioned out of freshly cut rushes.

After being allowed to stand at their sacred posts for two months, the *argei* were taken down and disposed of on May 15. Bayet (1969 97–98) opened up a new, important, and productive avenue for our understanding of these obscure rites by postulating that in early times there was a definite connection between the disposal of the *argei* on May 15 and the Roman observance of the Lemuria on May 9, 11, and 13. The latter were three days on which the early Romans believed that ghosts were about and needed to be appeased. Ovid (*Fasti* V.429–444) has preserved for us a vivid description of how the Romans propitiated potentially hostile spirits at this time. A person arose from bed at midnight. After washing his hands, making an apotropaic gesture, and wearing no shoes, he tossed black beans behind him while repeating nine times an exhortation for the spirits to receive the beans in place of himself and those of his household. Then after washing his hands again and clashing bronze vessels together, the person nine times urged the spirits to depart from the house.

Both Harmon (1978A 1457–1459) and Nagy (1985 10ff) have accepted the connection between the Lemuria and the disposal of the *argei* and have adduced such compelling arguments in its favor that we may now regard the thesis as having been demonstrated beyond doubt. The belief in ghostly visitations at this time of year may have stemmed from the fact that this was the season when the earth was most active in fostering the growth of crops, and that this activity was somehow thought to involve spirits of the nether world. May (mensis Maius) in fact was so named because this was a time of growth or "bigness." Alternatively, since the crops were now at a critical stage in their maturation (cf. the Robigalia of April 25 and the Ambarvalia celebrated at the end of May), it may have been thought that the crops at this time were especially in need of protection from unseen adverse forces, which in turn spawned the idea that ghosts were about and needed to be appeased. The *argei* stationed throughout the city, it seems likely, were intended to divert the attention of these chthonic spirits away from real people; and tossing them into the Tiber to be carried out to sea was designed to rid the community of the pollution absorbed by the *argei*.[6] Bayet regarded the removal of the rush puppets in mid-May as forming a lustral procession. The purgative character of the *argei* is asserted by Plutarch, who in item 86 of his *Roman Questions* terms their rites "the greatest of their [the Romans'] purifications." Dionysius of Halicarnassus states that the *argei* were bound hand and foot before being tossed into the Tiber. This was obviously done in order to prevent the rush dummies from doing any harm in case they were possessed by some evil spirit. Comparative ethnography provides similar examples of the popular belief in ghostly visitants coupled with rites of atonement, expulsion, and purification.[7]

A chthonic association with the *argei* is further suggested by the date March 16, the day on which the rites of the *argei* commenced, because this was an after day or black day (*dies postriduanus* or *dies ater*). As shown in the preceding chapter, the after days were ones on which it was forbidden to perform ceremonies requiring the invocation of Jupiter and Janus, Rome's paramount gods of primacy and firstness. Ovid, however, mentions that the procession to the chapels of the *argei* took place on both March 16 and 17. If, as is generally supposed, the ceremonies surrounding the *argei* were of great antiquity, conducting the procession to the chapels on two consecutive days would have been inconsistent with early Roman official religious practice: for nowhere else in the calendar do we find important ceremonies being assigned to two consecutive days. If rites could not be carried out on a single day, they were conducted on consecutive odd numbered days, as we have just seen in the case of the Lemuria. Consequently, Radke (1990B 7–8) has attempted to explain this anomaly in a novel fashion. After noting that the Umbrians, according to Roman writers, reckoned the day as beginning at noon, whereas the Romans, of course, did so from midnight, Radke constructs an elaborate scheme in which Umbrian diurnal time reckoning can be used to account for what Ovid has recorded, and at the same time an interval of an entire 24-hour day can be interposed in keeping with early Roman official religious practice. This could have been accomplished by having part of the ceremony performed before noon on March 16 and then not completed until after noon on March 17. Ingenious as this explanation is, it strikes this author as far too peculiar. Why would the Romans in this instance alone have had recourse to use an Umbrian scheme of time reckoning? Indeed, a far simpler and more satisfying explanation is readily at hand, but it has thus far been overlooked by modern scholars.

In Republican times, the interval between March 16 and May 15 (counting inclusively, of course, as the Romans always did) was exactly sixty days: 16 days from March 16 to 31, plus the 29 days of April, plus the 15 days of May. But when, as the result of the Julian reform of the calendar, April received a thirtieth day, a sixty-day interval counted backward from May 15 would have brought one to March 17, not March 16. It therefore seems obvious that the framers of the Julian calendar, wishing to maintain this sixty-day interval, ordained that henceforth the mid-March procession of the *argei* would be distributed over the two days of March 16 and 17 in order to preserve hallowed tradition along with a one-day shift caused by the calendrical innovation. We see exactly the same process in reference to the celebration of Augustus' birthday. When he was born on September 23 in 63 B.C., the month of his birth had only 29 days, and his birthday in Roman terms would have been expressed as "*VIII ante diem kal. Oct.* = eight days before the kalends of October." When, however, Rome's first future emperor was seventeen years old, the calendar was reformed, and September, like April, received a thirtieth day, so that *VIII ante diem kal. Oct.* was shifted by a day and became September 24, and September 23 had to be expressed as *IX ante diem kal. Oct.* Consequently, as we see from an inscription from Narbo, recording the establishment of an annual cult to the *Numen Augusti* (*ILS* 112), Augustus' birthday was to be celebrated on the two consecutive days, September 23 and 24 (*IX ante diem kal. Oct.* and *VIII ante diem kal. Oct.*). Thus, it is as if Roman religious experts reckoned the mid-March rites of the *argei* as occurring "*LX ante diem id. Mai.* = 60 days before the ides of May;" and when April was lengthened by a thirtieth day, the ceremony on March 16 was displaced by a day, and both the 16th and 17th of March henceforth were the temporal receptacles of the traditional ceremony. We should keep in mind that Ovid wrote his *Fasti* quite some time after the Julian reform of the calendar, about sixty years. This neat and simple explanation for the anomaly of March 16–17 gives us an interesting and important glimpse into the minds of the religious experts who participated in reforming the calendar at the end of the Republic, because it shows that they concerned themselves with how traditional religious practices were to be accommodated to the minor temporal displacements caused by the lengthening of the old months.

Why were there 27 *argei*? Their multitude argues against them being rain charms or offerings to appease the river god over whom the Romans had built a bridge, since such rites known from other cultures generally involve a single anthropomorphic figure.[8] Varro in his *De Lingua Latina* V.45–54 has preserved for us a partial list of the locations of the chapels of the *argei* throughout the city. His list derives from an official religious record and indicates that they were organized into four distinct groups, one for each of the urban tribes: the Suburana, Esquilina, Collina, and Palatina. Moreover, since Varro's partial list of 14 *argei* never assigns more than six chapels to any region of the city, some modern scholars have questioned the number 27 found only in Varro's text and have wished to emend the figure to 24, thereby allowing six chapels to be assigned to each region.[9] Alternatively, Richardson (1992 37–39) has recently suggested that the number should be 28 with each region having seven chapels. In addition, their stations have been used by some modern scholars to try to reconstruct Rome's early urban topography or its archaic civic organization.[10] Palmer (1970 84–97), for example, has associated the chapels of the *argei* with the curiae, archaic divisions of the Roman people. According to the ancient Roman tradition, these divisions of the populace numbered 30, but Palmer proposes that their number was increased over time as the early Roman state grew, and that the 27 *argei* represent a stage of development in which the community consisted of only 27 curiae. To be sure, there may have initially been some relationship between the number and/or location of these chapels and Rome's early urban topography and/or its civic organization, but the ordering of the chapels into four distinct groups may suggest that Varro's list reflects a long process of priestly reconfiguration of these sites. Thus, given the incompleteness of Varro's list and the strong possibility of later sacerdotal modifications, it seems best to accept Varro's number of 27 and to explore what significance that number might have.

The number of the *argei* is probably best understood as thrice nine.[11] Both three and nine are frequently encountered in Greek and Roman religion and magic. A provision in Rome's Law of the Twelve Tables, dating to c.450 B.C., specified that if a woman wished to avoid coming into the legal control of her husband, she had to interrupt each year of their cohabitation by absenting herself for three consecutive nights (Gaius *Institutes* I.111). Among various extraordinary religious measures designed to restore the goodwill of the gods (*pax deorum*) in the wake of C. Flaminius' disastrous defeat and death at the hands of Hannibal at Lake Trasimenus in 217 B.C., the Romans decided to celebrate games costing exactly 333,333 and one third *asses* (Livy XXII.10.7). Observing the Lemuria on three nights was certainly intended to augment the efficacy of the rites by the auspicious character of the number 3. The Roman state regularly expiated the portent of a rain of stones by an observance lasting nine days (thrice three), a ritual that they termed *sacrum novendiale* (= the nine-day rite).[12] As already noted, when placating the ghosts of the Lemuria, a person was supposed to repeat the words of the ceremony nine times. When Augustus celebrated the *ludi saeculares* in 17 B.C., a chorus of thrice nine boys and thrice nine girls sang Horace's *Carmen Saeculare* (*ILS* 5050 ll.147–149). As will be argued in Chapter 4, this ceremony was probably adapted from a religious procession of thrice nine maidens used for purifying the city and known to have been employed during the years 207–92 B.C.[13] It has gone unnoticed by all previous modern commentators that if a pious person performed the placatory rites on all three nights of the Lemuria, he would have repeated his exhortations thrice nine times.

It is noteworthy that these processions involving thrice nine maidens and the rites of the *argei* were both designed to purify the city, and both these rites of lustration were rendered even more efficacious by the virginity of the participants. Furthermore, in three of the seven lustral processions recorded in the ancient sources, the rites were performed by thrice nine maidens in the aftermath of civil bloodshed in the city: following the seditions of Ti. Gracchus, C. Gracchus, and L. Appuleius Saturninus. At these times, it may have been felt that the angry spirits of slain fellow citizens needed to be propitiated. In this regard it should be remembered that Ovid (*Fasti* V.445–484) explains the rites of the Lemuria as resulting from the need to appease the angry ghost of Remus, slain by his own brother when the two were founding Rome. Ovid's derivation of *lemures* (= ghosts) from Remus' name is clearly fanciful and is doubtless taken from earlier antiquarian literature, but the connection between the Lemuria and those who have died a violent death may likely represent the actual thinking of the early Romans, since it is a widespread popular belief that hostile ghosts are the spirits of those who have died before their time.[14] Thus, unless the similarities here outlined between the rites of the *argei* and the lustral processions were the product of later pontifical syncretism, the solution to the number of the *argei* would appear to lie in the belief in the efficacy of certain numbers rather than in vagaries of Rome's early urban or civic development.

According to the rules of Latin grammar the nominative singular of *argei* should be *argeus*, a form which not surprisingly is unattested in the extremely exiguous ancient literary evidence concerning these rush puppets. The -eus ending is not uncommonly employed in Latin to generate adjectives by adding the suffix to the base of a noun; but the element is encountered as a termination of nouns as well.[15] Thus, the term *argei* consists simply of the two elements *arg-* and *-eus*. Furthermore, the adjective *scirpeus* meaning 'of rushes' possesses this same suffix and is closely associated with the term *argei* in the ancient antiquarian literature; for Varro, Ovid, and Festus use this word to indicate that the *argei* were composed of rushes.[16] Both words' trisyllabic nature and their sharing of the same termination could indicate that the close association of *scirpeus* with *argei* led the early Romans to attach the *-eus* suffix to the etymon *arg-*. If so, this would constitute very early evidence that the *argei* were in fact fashioned out of rushes. The point is not a trivial one, because Holland (1961 313ff) argued that the *argei* were not composed of rushes but of the straw left over from the Vestals' making of *mola salsa* during the period May 7–14. She pointed out that Ovid (*Fasti* V.631) uses the adjective *stramineus* (= of straw) in reference to the *argei*, but this is the only such usage. More importantly, Ovid begins and ends his account of the *argei* by describing them as composed of rushes. Consequently, given the looseness of poetic language, it seems better to conclude that rushes were used to make the human-like figures. Since no folk beliefs are encountered in the relatively infrequent ancient references to rushes, the Romans' choice of material may have stemmed from practical considerations. Durability and buoyancy would have been required of objects left out in the open for two months and then tossed into the Tiber to be carried out to sea; and the ancients recognized these qualities in rushes as seen from their use in the construction of items of farm equipment and small sailing craft.[17]

Holland, however, drew scholars' attention to the role of the Vestals in tossing the rush dummies into the Tiber and the fact that this rite came immediately after they had spent several days in making *mola salsa*, a salted meal made from the season's first ears of grain and used to sprinkle on the heads of sacrificial victims. Nevertheless, Holland's notion that the anthropomorphic figures were merely the product of the Vestals' doodling with left-over straw is too simplistic and falls into a common error made by modern scholars of early Roman religious festivals: namely, the tendency to try to encompass a complex of rites into a single explanation. The Lemuria and the making of *mola salsa* may have occurred at the same time not due to any direct connection but simply resulting from their mutual association with the growth of crops in May. In addition, the fact that the Vestals were required to discard the by-products of their activities could explain why they were designated by the pontiffs to throw the *argei* into the Tiber. In doing so, the pontiffs merely combined two purgative rites into one.

But to return to the etymology of *argei*, most scholars since Wissowa have been content to accept the word's derivation from Greek *Argeioi*. For example, Maddoli (1971) has argued that tossing the *argei* into the Tiber is a Roman adaptation of a Greek religious practice: the ritual bathing of the cult statue of Argive Hera, introduced into Rome during the sixth century B.C. from the Greeks of Magna Graecia through Etruscan mediation. Nagy (1985 17–20), on the other hand, accepts the basic historicity of the later ancient tradition that viewed the *argei* as substitutes for actual Greeks (*Argeioi*) offered up by the Romans in human sacrifice. Like Maddoli, Nagy regards the Etruscans of the sixth century B.C. as having played a pivotal role in the Roman adoption of these rites. This author, however, regards this line of modern interpretation as totally misguided. Rather, in his opinion, Fowler long ago (1899 113 and 118) had it right, and we would do well to follow his lead.

The element *arg-* present in *argei* is also encountered in the Latin words *argilla* (= white clay) and *argentum* (= silver).[18] Moreover, the latter has cognates in Greek *argyrion* and Old Irish *argat*. Thus, the term *argei* may have originally described something or someone of a white or silvery color. Compare Greek *argos* = 'shiny', Sanskrit *arjuna* = 'white' or 'light', and Hittite *harki* = 'white'. W. Mannhardt in his *Baumkultus* of 1875, catalogued May Day and Whitsuntide celebrations attested throughout early Europe involving the immersion into water of a person decked out in green foliage or of a human-like figure fashioned out of greenery.[19] In these and related rites it was not uncommon for the celebrants to refer to the person or object as "the old man" or "the old woman." Thus, the term *argei* must have originally meant something like "the hoary-haired ones." Note the Romans' use of the adjective *canus* (= white) to mean 'old man'. The appellation would have suited rush dummies that had lost much of their verdant freshness over the course of two months. Moreover, we may plausibly speculate that when the Romans began to forget that *argei* meant "the aged ones," someone or someones (perhaps priests who were still in the know) must have coined another expression to replace it: "*sexagenarios de ponte*." This brings us back to the sixty-day interval discussed above. The fact that the rush dummies remained at their stations for exactly sixty days explains why the Romans chose to call them *sexagenarios* rather than *septuagenarios*, *octogenarios*, or *nonagenarios*. Roman antiquarians of the late Republic and early Empire, however, baffled by the term *argei* and how it was supposed to be connected with the expression "*sexagenarios de ponte*," contrived fanciful explanations for both these linguistic curiosities. On the one hand, these later writers equated *argei* with Greek *Argeioi* through the myth of Hercules coming to Rome and initiating the civilizing process of Hellenization by ending the barbarous aboriginal practice of human sacrifice. On the other hand, they devised an alternative explanation for "*sexagenarios de ponte*" involving an otherwise unattested event or practice of olden times, according to which the younger citizens forced their elders off the voting ramps, also called *pontes*, used in Roman assemblies (Ovid *Fasti* V.633–634 and Festus 452L). The political violence that frequently disturbed public meetings during the late Republic must have rendered this alternative view far more meaningful to contemporary Romans than an explanation involving an outmoded and poorly understood religious ritual. Yet, despite its fanciful character and the fact that it is recorded only in association with the expression "*sexagenarios de ponte*," this "comitial" interpretation of these words has sometimes found its way into modern scholarly treatments of Roman assemblies and voting procedures.[20]

Finally, Mannhardt in his *Wald und Feldkulte* (1877 265–273) attempted to explain the rites of the *argei* in mid May in terms of other early peoples' rituals involving the death and/or revival of a spirit of vegetation by immersion into water, but as he well realized, mid May seems to be somewhat early for such a ceremony. Nevertheless, his association of the *argei* with other early European practices was endorsed by Fowler and Rose.[21] Frazer, however, in his commentary on Ovid's *Fasti* (IV. 83–85) objected to Mannhardt's thesis not only on the grounds of the early date but also because the rushes from which the *argei* were made do not suggest that the figures were viewed in any way as embodying fertility. Frazer's latter point is quite valid, but the parallels adduced by Mannhardt are equally compelling. Thus, we may be justified in concluding that the Romans adapted a widespread primitive custom to their own ends. Human-like figures were constructed out of fresh greenery and set up in mid March to symbolize the reawakening of plant life, but having them remain on display throughout the city over the next two months until the Lemuria served the additional function of purifying the community of harmful spirits during a critical time of the crops' growth. Moreover, just as other peoples commemorated the waning of the spirit of vegetation by referring to an anthropomorphic object as "the old man," so the Romans retained the concept of old age in the term *argei* and in the expression "*sexagenarios de ponte*" even to the extent that the sexagenarian senility of the *argei* was made to correspond to a sixty-day interval in the calendar.

4 Origin and History of the *Ludi Saeculares*[1]

All of the festivals discussed in the preceding chapters were celebrated by the Romans annually. We now come to one of the most peculiar Roman celebrations, the *Ludi Saeculares*, observed not annually but every *saeculum*. Thanks to the discovery of the epigraphic texts of the official *Acta* of their celebration from Augustus to Septimius Severus, our knowledge of these games during Imperial times is fairly detailed, but the converse is true for their origin and history during the Republic. Nevertheless, an examination of the history of this celebration, spanning roughly six centuries, can reveal much about how Romans adapted religious rites to fit the changing needs of their society.

THE CHARTER MYTH

"Because it is obvious from their very names whence the other games derive, it does not seem absurd to recount the origin of the saecular games,[2] whose nature is less well known. When the city and its fields were being devastated by a great plague, Valesius, a rich man of a rustic temperament, had two sons and a daughter who were suffering to the utter despair of the physicians. While fetching hot water for them from the hearth, he knelt down on his knees and besought the household Lares to transfer his children's danger onto his own head. A voice then came forth and said that he would save them if he carried them to Tarentum, conveyed straight along by the Tiber River, and if he refreshed them there with water taken from the altar of Father Dis and Proserpina. Although he was much confused by this prediction because a long and dangerous voyage was being commanded, doubtful hope nevertheless overcame his present fear, and he immediately carried his children down to the bank of the Tiber; for he lived in a villa near the village of Eretum in the Sabine territory. While heading for Ostia under sail, he put into land at the Campus Martius at bedtime. He wished to relieve the thirst of the sick, but since fire was not available on shipboard, he learned from the helmsman that smoke was seen not far away. After being ordered by him to set out for Tarentum (for that was the place's name), he eagerly seized a cup, filled it with water from the river, and carried it, more joyously now, to the place from which the smoke had arisen, thinking that he had obtained close by some traces of the divinely granted remedy. The ground was smoking more than just having the remains of a fire. Seizing even more upon the omen, he scraped together light kindling and whatever material happened to be at hand; and by constant blowing he brought forth a flame. He then gave the heated water to the children to drink. After drinking, they fell into a healthy sleep and were instantly freed from the persistent grip of the illness. They informed their father that in their dreams they had seen their bodies being wiped off with a sponge by some god, and that they were instructed to hold lectisternia and nocturnal games and to sacrifice black victims at the altar of Father Dis and Proserpina from which the drink had been brought to them. Since he had seen no altar at the place, he believed that it was desired that he erect one. He proceeded to the city to purchase an altar, having left behind ones who were to excavate the earth down to bedrock for laying the foundations. Following their master's orders, they reached a depth of 20 feet in their digging and came upon an altar inscribed 'to Father Dis and Proserpina'. Valesius, on learning this from his slave messenger, abandoned his plan to purchase an altar and sacrificed at Terentum black victims that in ancient times were called *furvae*. He also staged games and *lectisternia* on three consecutive nights, because the same number of children had been freed from danger. Valerius Publicola, who was first consul, followed this man's precedent when wishing to help his fellow citizens. After vows had been publicly pronounced at the same altar, after black cattle had been slain (males to Dis and females to Proserpina), and after a *lectisternium* and games had been held for three nights, he covered the altar with earth, as it had been before."[3]

Valerius Maximus, writing during the reign of Tiberius, records this tale in the second book of his *Memorable Deeds and Sayings*. It was clearly designed to serve as the charter myth for the *Ludi Saeculares* with each element in the story providing an explanation for each major facet of the celebration. For example, the story accounts for the location of the games at the Tarentum (or Terentum) in the Campus Martius, the rites being performed for three nights in honor of Father Dis and Proserpina, the existence of an underground altar to these divinities, and the games being accompanied by a *lectisternium* (actually a *sellisternium*). In addition, the tale is cast in the form of a sacred story, resembling the well-known Greek historical *genre* of the foundation tale (*ktisis*) for a colony. The story begins with a crisis, in this case a plague. In a Greek foundation tale we would next expect the people of the stricken community to consult an oracle, such as Delphi; and after being given an enigmatic response, they would set off in search of a place to colonize, and during the course of their travels something peculiar would happen to fulfill the prophecy in a surprising fashion and would convince them that they had in fact found the place predicted for them by the oracle. In this case, however, a particular individual seeks divine assistance from his household gods and is given a riddling response: to sail along the Tiber River to Tarentum and to cure the three afflicted children by giving them water heated upon the altar of Father Dis and Proserpina. In the manner of a Greek colonial foundation story, Valesius thinks that he is supposed to seek out Tarentum, the well-known Greek city in southern Italy, but he succeeds in stumbling upon the true meaning of the prophecy when he happens upon a similarly named place downstream along the Tiber from Eretum in the Campus Martius at the site of Rome. In the somewhat more detailed version of the story recorded by Zosimus (II.1–2) the puzzling nature of the divine advice is made clearer, because Valesius is urged to do what seems impossible: to sail to Tarentum, to draw water from the Tiber, and to heat it on the hearth of Dis and Proserpina.

We may plumb the story a bit further in search of the original idea behind the *Ludi Saeculares*. Since the Sabine Valesius is the first to perform these rites in gratitude for his children recovering their health from a serious illness, we may suppose that the Saecular Games were originally thought to protect the Roman population from the visitation of epidemic diseases. The conjecture obtains support from Valesius' very name, because he was doubtless chosen to be the central figure in this charter myth due to his name's resemblance to the Latin word for health (*valetudo*) and the verb for being well (*valere*). The etymological connection is reinforced by the final statement in the passage, in which Valerius Publicola is said to have employed these same rites at the beginning of the Republic to assist his fellow citizens. Plutarch (*Publ.* 21.1) provides us with further information on this point. He writes that during Publicola's fourth consulship in 504 B.C. the women of Rome were visited with a plague of miscarriages, and that the crisis did not pass until Publicola had consulted the Sibylline Books and had revived rites to Hades.[4] Both Livy (II.16.2–6) and Dionysius (V.40–43) do not record this incident in their narratives of this year. Plutarch, however, at the very beginning of his *Life of Publicola* says that this man was descended from the Valerius who had made peace among the Romans. Dionysius (II.46.3) explicates this vague assertion by placing it in the context of the war between the Romans and Sabines at the beginning of Romulus' reign, ending in peace and the joint rule of Romulus and T. Tatius. Dionysius says that three leading Sabines decided to settle in Rome at this time: Volusus Valerius, Tallus Turannius, and Mettius Curtius. It therefore seems likely that the Valesius in the charter myth, who lived at Eretum among the Sabines, is to be identified with this Valerius of Romulus' reign and the ancestor to Publicola, who in turn was regarded as the founder of the great patrician family of the Valerii. Dionysius' Volusus Valerius is doubtless the same as Plutarch's Velesus, a Sabine settled at Rome and sent out at the end of Romulus' reign to persuade Numa to leave Sabine Cures to become Rome's next king (Plut. *Numa* 5.1).

Much of this Valerian material, as Wiseman (1998 75–89 and 165–167) has cogently argued, must have been enhanced (if not actually invented) by Valerius Antias during the closing years of the Republic. In fact, in his discussion of the *Saeculum* and the Saecular Games Censorinus (17.8–11) cites Valerius Antias four times in dating the first, second, third, and fourth celebrations of the *Ludi Saeculares*. It is therefore quite likely that Antias is Valerius Maximus' source for the charter myth.[5] Coarelli (1993 214–229), on the other hand, explains the Valerian connection to the Saecular Games both in historical and historiographical terms. Although he sees the hand of Valerius Antias in shaping the chronology of the *Ludi Saeculares* in Republican times, he also regards Publicola's association with the rites as historical. He argues that the Tarentine rites were initially observed only by members of the Valerian family until they were taken over by the state in 249 B.C. in a manner similar to the cult of Hercules at the *Ara Maxima*, which, according to later tradition, was at first in the hands of two families, the Pinarii and Potitii, until the cult was absorbed by the Roman state in 312 B.C. Bernstein (1998 135–142) has adopted a similar view, arguing that the rites at the Tarentum were initially private ones of the Valerian family until in 249 B.C. consultation of the Sibylline Books caused them to be Hellenized and taken over by the state as the *Ludi Saeculares*. Finally, before leaving the murky *primordia* of the Saecular Games, we should mention another Valerian association with the abatement of disease in that collection of bizarre tales, the pseudo-Plutarchian *Parallela Minora* (35), according to which Valeria Luperca cured the sick of Falerii by tapping them lightly with a magical hammer.

WHAT WAS A *SAECULUM*?

Thanks to the seventeenth chapter of Censorinus' *De Die Natali*, we know what a *saeculum* was, and that it originated with the Etruscans. In mere human terms, a *saeculum* was simply the longest possible life span; but when applied to the history of communities or entire peoples, it took on a different meaning. Censorinus (17.5–6) explains that the Etruscan nation had been allotted ten *saecula* by the gods, and that Varro, apparently his source of information, was writing during the eighth Etruscan *saeculum*. The first *saeculum* of a people begins with the founding of their community and ends when the last person among the original founding population dies. At that point, the second *saeculum* begins and ends with the death of the last surviving person among those who happened to be alive at the beginning of the second *saeculum*. Censorinus further records the lengths of the preceding seven Etruscan *saecula* as follows: the first four each lasted 105 years, the fifth 123, and the sixth and seventh each endured for 119 years, adding up to a total of 781 years. Van Son (1963 270–273) has argued convincingly that the seventh Etruscan saeculum began in 207 and ended in 88 B.C. Thus, according to this Etruscan scheme, their nation had come into being c.869 B.C.; Varro was alive and writing during the eighth *saeculum* (post 88 B.C.); and the Etruscan people were therefore doomed to pass out of existence at some time during the second century of our era.

At some point the Romans borrowed this notion of the *saeculum* from the Etruscans and wove it into their ideas of the *Ludi Saeculares*. They were supposed to be celebrated only once every 100 years or so, so that no one could ever witness them twice. Although our information about the Roman concept of the *saeculum* and of the Saecular Games during Republican times is very meager, we possess several important *testimonia* confirming unequivocally that the Romans of the late Republic regarded the *saeculum* as lasting exactly 100 years and thus corresponding to our modern notion of a century. The earliest of these *testimonia* is a fragment of L. Calpurnius Piso Frugi, a verbatim quotation preserved by Censorinus himself (17.13) and quoted to show that Republican writers did in fact treat the *saeculum* as consisting of exactly 100 years. In fact, Piso used the term just as we today use centuries in calculating the passage of time in periods of 100 years: for in giving a simple date formula in his narrative he wrote that Rome began its seventh *saeculum* (i.e., century) with the consulship of 158 B.C. The fragment not only shows that the Romans of Piso's day regarded the *saeculum* as a century of time, but it further demonstrates that the historians were also using the idea in establishing Rome's foundation date and their chronologies *ab urbe condita*. Secondly, Censorinus (17.8) states that according to "Antias and other historians" the Saecular Games were supposed to be observed every 100 years. Indeed, Censorinus' citation of Valerius Antias in the following three sections shows that Antias dated the *Ludi Saeculares* to the years 348, 249, and 149 B.C.[6] A third *testimonium* concerning the Republican length of the *saeculum* is a brief statement in Varro's *De Lingua Latina* VI.11, occurring in a passage in which he is explaining terms relating to periods of time: "they call the space of 100 years a *seclum*, so-called from old man (*sene*), because they thought it to be the longest period of people growing old."[7] It is evident that Varro has used a slightly different spelling of the word in order to make more plausible his etymology: *seclum* < *senex*. Yet, it is important to note that Varro wrote this statement c.45 B.C., about a generation before Augustus' celebration of the *Ludi Saeculares* in 17 B.C. Censorinus (17.8) confirms this Varronian judgment concerning the length of the *saeculum* by supplying us with a verbatim quotation from the first book of Varro's *De Scaenicis Originibus*. The last testimonium again comes from Censorinus (17.9) and is a verbatim quotation from Livy's lost 136th Book in which the historian recorded Augustus' celebration of the Saecular Games, which, Livy noted, were traditionally observed every 100 years.

Following his explanation of the Etruscan *saeculum*, Censorinus provides us with a chronology of the Saecular Games for both the Republic and the Empire. Moreover, since we have other corroborating literary sources and the epigraphic texts of the *Acta* themselves, the chronology for the Saecular Games in Imperial times is not in doubt: celebrated in 17 B.C. by Augustus, in 47 A.D. by Claudius to coincide with the 800th year of Rome's existence (Tac. *Ann.* XI.11),[8] in 88 A.D. by Domitian 104 years after those of Augustus (Suet. *Dom.* 4.3 and Tac. *Ann.* XI.11), and in 204 A.D. by Septimius Severus 220 years after those of Augustus. Censorinus quotes the *Carmen Saeculare* composed by Horace for the Augustan celebration in support of an alternative notion that the *saeculum* was a period of 110 years. Modern scholars generally agree that this alternative view was of Augustan origin and was designed to justify Augustus' desire to celebrate these games in 17 B.C. to inaugurate the establishment of his principate. Evidence for Augustan mendacious tampering with the earlier history of the Saecular Games to fit his 110-year scheme is obvious from Censorinus, who cites the *Commentarii* of the *quindecimviri sacris faciundis*, the priests in charge of administering the games, for the dates of their celebration in Republican times; and it is not surprising to find that according to this chronology, they were observed at exactly 110-year intervals: 126 B.C., 236 B.C., 346 B.C., and 456 B.C. Thus, by the beginning of the principate there existed two competing notions of the *saeculum*'s length, 100 vs. 110 years; and this difference in opinion was exploited by emperors in order to grant them the honor of celebrating these rarest of Roman games. Claudius followed the 100-year tradition of the Republic and thus wound up observing the Saecular Games a mere 63 years after those of Augustus, whereas Domitian dated his celebration from Augustus and held the games just 41 years after those of Claudius. Septimius Severus was able to rely upon the 110-year notion and thus celebrated the games exactly two Augustan *saecula* after those of Augustus himself.

WHEN WERE THESE RITES FIRST OBSERVED?

During the winter of 1886/7, as the result of the construction of the Corso Vittorio Emanuele, the remains of an ancient altar, thought to be the Tarentum, was uncovered in the Campus Martius, sixteen feet below the modern ground level.[9] The remains consisted of two large stone blocks. The altar is estimated to have measured about eleven feet square, was placed upon a podium, and was approached by three steps. Then nearly four years later, in September of 1890, as workmen were engaged in digging a sewer associated with the *Lungotevere* project, twelve inscribed marble fragments were found embedded in a Medieval wall, located about 300 yards north of the supposed site of the Tarentum. The inscriptions recorded the official *acta* of the Imperial celebrations of the *Ludi Saeculares*. One small fragment came from Claudius' reign, and the other eleven were Severan, most of which were quite fragmentary. They were published in *CIL* VI as 32,325–32,336 together with two other epigraphic texts of the Augustan *Acta* discovered in the sixteenth century, the latter being items 32,323–32,324. The former (32,323) was a very informative, substantial text of 167 lines, whereas the latter (32,324) was rather small and fragmentary. Although Mommsen regarded 32,324 as belonging to Claudius, Moretti (1985) has now demonstrated that it formed part of the Augustan *Acta*. Dessau subsequently printed the Augustan and Severan texts (with some omissions) as *ILS* 5050 and 5050A respectively. Then in 1930 a new fragment of the Severan *Acta* (*AE* 1932 #70) was discovered near the site of the other fragments. Consequently, the modern scholarly consensus is that the Tarentum is not the site of the ancient altar unearthed in 1886/7, but it is to be situated in the vicinity of the epigraphic finds, inside the northwestern bend of the Tiber near the Ponte Sant'Angelo (Pons Aelius) and the Ponte Vittorio Emanuele.[10]

The fragment of the Severan Acta discovered in 1930 rekindled scholarly interest in the early history of the games. What especially caught the attention of scholars was part of a prayer to Apollo: "quaeso precorque uti tu imperium maiestatemque p(opuli) R(omani) Q(uiritium) duelli domique auxis, utique semper Latinus obtemperassit = I ask and pray that you increase the power and greatness of the Roman people the Quirites in war and at home, and that the Latin forever be obedient." Taylor (1934) argued persuasively that the prayer to keep the Latin obedient hearkened back to the fourth century B.C. before the dissolution of the Latin League in 338 B.C., and thus it established a probable *terminus ante quem* for the earliest observance of the Saecular Games. Moreover, since the celebration always involved a *lectisternium* and stage performances, these two features, first introduced into Rome respectively in 399 and 364 B.C., also constituted in Taylor's view *termini post quos*. Taylor found in Livy's brief recording of a plague, consultation of the Sibylline Books, and observance of a *lectisternium* in 348 B.C. (VII.27.1) the most likely time when rites were observed at the Tarentum to placate Father Dis and Proserpina. Furthermore, following a line of reasoning developed by Roth (1853 373–374), she argued that the rites were not established as centennial until they were revived in 249 B.C., from which time onwards they were properly known as *Ludi Saeculares*. Taylor also suggested that in order to maintain the notion of a 100-year *saeculum*, Valerius Antias contrived the four fictitious dictator years (333, 324, 309, and 301 B.C.) that were later incorporated into the Attican and Varronian chronologies of early Rome, which became the standard for dating *ab urbe condita* from Augustus onwards.

The great L. R. Taylor was certainly on the right track in trying to locate in the fourth century B.C. the first celebration of the ceremonies that later became the *Ludi Saeculares*, but in the author's view she erred in assigning the first observance to the year 348 B.C. A more likely context would appear to be 362 B.C. The plague and *lectisternium* of 348 receive the barest attention in Livy's narrative, whereas his account of the years 364–362 is much more elaborate and suggests a crisis far more severe and thus more suitable for a major religious innovation. According to Livy (VII.2), when a plague broke out in 364, the Romans dealt with it first by observing the third *lectisternium*, but when the plague persisted, the Romans for the first time, among other *placamina*, instituted public stage performances (*ludi scaenici*) introduced from Etruria. This innovation prompts Livy to devote the remainder of the chapter to a digression on the early history of Roman drama. When Livy resumes his narrative at VII.3.1, he reports that the plague continued into the next year (363 B.C.) accompanied by the flooding of the Tiber. When other remedies of assuaging divine anger were sought out, recourse was finally taken in the appointment of L. Manlius as dictator for driving the nail. This then leads Livy to digress slightly to discuss the possible meaning of this custom. His next two chapters (VII.4–5) begin his narrative for the following year (362 B.C.) and describe the unsuccessful attempt on the part of a plebeian tribune to prosecute L. Manlius for harsh conduct as dictator. Then in VII.6.1–6 Livy tells the strange story of a chasm opening up in the middle of the Forum; and after people unsuccessfully tried filling it up by throwing into it large quantities of earth (*coniectu terrae*), seers (*vates*) declared that the Romans must make an offering of the one thing that made the Roman people most powerful, because this would ensure the continuance of the state. Whereupon M. Curtius, surmising that Rome's strength lay most in its brave soldiery, armed himself and rode into the chasm on his horse, and his self sacrifice was accompanied by many men and women making their own offerings of gifts and fruits into the chasm (*donaque ac fruges*), which then closed as mysteriously as it had opened. Livy ends the story by remarking that this tale accounts for the Lacus Curtius in the middle of the Roman Forum more convincingly than the story told of Mettius Curtius at the time of the battle between the Romans under Romulus and the Sabines under T. Tatius.

What are we to make of this, and how might it relate to the origin of the *Ludi Saeculares*? It looks as if during the three years 364–362 B.C. Rome was afflicted first by a very persistent plague, followed by the flooding of the Tiber. The latter, no doubt, overflowed into the Forum and totally submerged the low area of the Lacus Curtius. If the Romans at this time had attempted to appease the nether gods by opening the earth at Tarentum (or Terentum = the earth place)[11] in the Campus Martius, where they made offerings to Father Dis and Proserpina, we can understand how a historian centuries later might conflate the flooding of the Lacus Curtius and the solemn rites at the Tarentum into the fanciful story of Curtius' *devotio* and the explanation for the toponym Lacus Curtius. The original meaning of the latter toponym was probably "The Fenced-in Pool," but as the result of the emergence of the Curtian family into Roman public affairs toward the close of the second century B.C. antiquarian speculation reinterpreted the toponym Lacus Curtius as having derived from an ancient member of this family.[12] Varro (*L. L.* V.148–150) indicates that there were current three different explanations for the Lacus Curtius: (1) from the Sabine Mettius Curtius at the time of King Romulus, recorded by L. Calpurnius Piso; (2) from the consul of 445 B.C. (the only member of the Curtian family ever to have reached the consulship in Republican times), recorded by Q. Lutatius Catulus; and (3) from M. Curtius, who devoted himself and rode on his horse into the chasm at this spot in the Forum. Varro ascribes this last version to Procilius, about whom virtually nothing is known; but since aspects of his tale resemble a prodigy mentioned by Plutarch in connection with Sulla at the time of the Social War (Plut. *Sulla* 6.6–7 with Forsythe 1994 157), this third explanation is likely to have come into being around this time.[13]

Livy's immediate source for this Curtian legend was most likely Q. Claudius Quadrigarius, who could have borrowed the story from Procilius and thus introduced it into the annalistic tradition (see the similar accounts of Dion. Hal. XIV.11 and Val. Max. V.6.2). Unlike his annalistic predecessors, whose histories began with Rome's remotest mythical origins, Quadrigarius commenced his account of Roman affairs with the year 390 B.C., apparently on the grounds that the Gauls had destroyed all public records, thus depriving earlier Roman traditions of any valid evidentiary basis.[14] The numerous fragments of Quadrigarius clearly display a relative lack of interest in domestic politics, but a very keen interest in military matters, especially of exploits of personal bravery.[15] The self immolation of the brave warrior M. Curtius thus fits well with what the fragments tell us of Quadrigarius' predilections. In addition, Livy's rather polemical tone in VII.6.5–6 in dismissing the explanation associated with Mettius Curtius is likely to be Livy's own reworking of similar remarks made by Quadrigarius, who wished to replace the Curtian tale from the mythical times of Romulus' reign with one dating to a much later, and hence, more credible period.

Assigning the first major performance of rites at the Tarentum to the year 362 rather than to 348 B.C., as Taylor proposed, has the advantage of accounting for two curious chronological puzzles. When we remove the four fictitious dictator years from the consular list for this period, the Varronian year 362 become the absolute date 358 B.C.[16] As will be treated below, the rites at the Tarentum were performed again in 249 B.C., at which time they were probably for the first time regarded as saecular, i.e., needing to be repeated every 100 years. The year 249 B.C. was the 110th year after 362/358 B.C. and therefore could have been the one solitary, historically valid datum seized upon by the *quindecimviri sacris faciundis* of Augustus' day to justify their notion of a 110-year *saeculum*. Secondly, if some Romans in later times knew that the rites at the Tarentum had been observed in 362/358 B.C., and that they had been ten years overdue when repeated in 249 B.C., it could have prompted them to reconstruct early Roman chronology by counting back several centuries from 358 B.C. This would neatly explain a fragment of Calpurnius Piso quoted verbatim by Censorinus (17.13), according to which the annalist wrote that in the 600th year after Rome's foundation the seventh *saeculum* began with the consulship of M. Aemilius Lepidus and C. Popillius for the second time (i.e., 158 B.C.).[17]

THE CALENDRICAL DATE OF THE *LUDI SAECULARES*

Both the Augustan and Severan *Acta* show that the Saecular Games were always observed during the first three days of June. When viewed in terms of the month's religious associations discussed in Chapter 1, it becomes obvious that Roman priests assigned the *Ludi Saeculares* to these three days quite deliberately. Performing the rites over a three-day period was clearly designed to increase their religious efficacy through the supposed felicity of the number three; and performing the ceremonies during the night was consonant with the worship of divinities of the underworld, to whom black animals were also appropriately sacrificed. Juno was originally the goddess of youthful vigor. Her month coincided with the maturation of the crops, and the Matralia of June 11 was concerned with the ripening or maturation of young women. Not only was June Juno's month, but June 1 was the most Junonian day of the year, because the first day of each month was sacred to Juno. As we have already seen, June 1 was notable for its rites to Carna, a divinity that protected one's internal organs. This would have been a perfect day on which to begin the observance of rites intended to protect all Romans from epidemic diseases.

Another perspective from which we can analyze the purpose and placement of the *Ludi Saeculares* in June is by examining the *Ludi Taurei*. Even though the latter are mentioned briefly only four times in literary texts and once in an inscription, what little we do know about them suggests that both the *Ludi Taurei* and the *Ludi Saeculares* were in part kindred outgrowths of the same Roman religious mentality. A very mutilated entry in Festus (478L s.v. *Tauri Ludi*) assigns the origin of these games to the reign of Tarquinius Superbus and explains that they were designed to placate the nether gods, because an epidemic of miscarriages was afflicting pregnant Roman women as the result of them having eaten the meat of sacrificed *tauri*. Servius Auctus, commenting on Vergil's *Aeneid* II.140, writes as follows: "If it [the cow to be sacrificed] were pregnant, it is called a *forda*, but if it is sterile, it is termed *taurica*, whence the *Ludi Taurei* take their name. They were established by King Tarquinius Superbus in accordance with the Sibylline Books, because every childbirth of women miscarried. Others say that the Ludi Taurei were established by the Sabines because of a plague, so that the public blight might be turned upon these sacrificial victims."[18] These two divergent explanations of the games' creation roughly correspond to two traditions associated with the origin of the Saecular Games. Servius Auctus' mention of a pestilence among the Sabines resembles the charter myth of the Ludi Saeculares quoted at the beginning of this chapter; whereas Plutarch's brief mention of Publicola reviving rites to Hades to avert a plague of miscarriages in Rome agrees very closely with the other explanation, except that it is dated to the reign of Tarquinius Superbus. Moreover, if Festus' *tauri* are understood to be steers or barren cows, we can easily understand through the principle of sympathetic magic why the Romans would have believed that ingestion of their meat by pregnant women would have resulted in miscarriages. Thus, the *Ludi Taurei* must have been observed in order to ward off barrenness; and just as the Fordicidia of April 15 centered around the sacrifice of pregnant cattle to promote the fertility of the growing crops, so the victims of the *Ludi Taurei* must have been barren and were intended to remove all barrenness from the Roman population.

Whenever the Romans founded a colony or extended the circuit of Rome's *pomerium*, the following ritual was used to mark out this sacred boundary. A fertile cow (*vacca*) and a steer (*taurus*) were yoked together and used to plow a furrow that then became the community's sacred boundary. The cow was kept on the inside, while the steer walked on the outside, because the Romans obviously wished to enclose the settlement with fertility and to banish sterility to the outside.[19] Thanks to a brief statement of Varro (*L. L.* V.154) giving an etymology for the Circus Flaminius, we know that the *Ludi Taurei* involved horse races in this part of the Campus Martius: "Also, for a similar reason the Circus Flaminius is so called, which is built around (*circum*) the Campus Flaminius, and because there too horses in the *Ludi Taurei* run around (*circum*) the turning posts."[20] Since this area lay outside the *pomerium*, it is quite obvious that the *Ludi Taurei* were performed here in order to keep barrenness outside of Rome's sacred boundary. The wording in the mutilated entry in Festus, as restored by Mueller and followed by Lindsay, hints at the same thing: "they were performed . . . so that the nether gods might not be summoned inside the walls = fiunt . . . [ne] intra muros evocentur d[i inferi]. . . ."

The remaining two ancient *testimonia* concerning the *Ludi Taurei* add other details of interest for the Saecular Games. Livy at XXIX.22.1 simply writes: "During those days in which these things were reported from Spain, the Ludi Taurii were performed for two days for the sake of religion = Per eos dies, quibus haec ex Hispania nuntiata sunt, Ludi Taurii per biduum facti religionis causa." The brief notice comes toward the end of his treatment of the year 186 B.C. infamous for the Bacchanalian affair that involved charges of murder through poisoning (Livy XXXIX.8–13 and 17–18, cf. Polyb. VI.13.4). Moreover, for the previous year Livy (XXXVIII.44.7) records: "then in accordance with the decree of the *decemviri* there was a *supplicatio* for three days for the health of the people, because a serious plague was ravaging the city and countryside = Supplicatio inde ex decemvirorum decreto pro valetudine populi per triduum fuit, quia gravis pestilentia urbem atque agros vastabat." Thus, the observance of the *Ludi Taurei* in 186 B.C. might have been due in part to this epidemic. Finally, the *Fasti Ostienses* for the year 145 A.D. (*Inscr. Ital.* VIII. I. 205) record the performance of *Ludi Taurei Quinquennales* over two days, June 25–26. The two day period agrees with Livy's brief notice recorded for the *Ludi Taurei* performed 330 years earlier. The absence of these games in the epigraphic calendars of the early principate might be due to the fact that they were quinquennial, not annual. Their quinquennial nature is clearly reminiscent of the Republican practice of the censors performing a lustration of the Roman people at the end of their censorship; and it is noteworthy that the Saecular Games, the *Ludi Taurei*, and the censorial lustration were all performed in the Campus Martius outside the *pomerium*. All three ceremonies doubtless shared the same basic purpose of purifying the Roman people and protecting them from unseen harmful forces, and they all did so at fixed temporal intervals. Finally, not only are the ancient aetiologies for the Saecular Games and the *Ludi Taurei* similar, but both ceremonies were performed in June. Thus, there exists the possibility that the *Ludi Taurei* may have formed the religious framework out of which the Romans fashioned the notion of the Saecular Games.

FROM *LUDI TARENTINI* TO *LUDI SAECULARES*

"Valerius [Verrius] Flaccus records that the Sacular Hymn and sacrifice were established every 110 years to Dis and Proserpina during the First Punic War in accordance with the reply of the *decemviri* after they had been ordered to inspect the Sibylline Books because of a prodigy that had occurred in that war: for part of the city wall was struck by lightning and collapsed. They therefore replied that the war against the Carthaginians could be waged successfully if games were celebrated and a hymn sung amid sacrifices in honor of Dis and Proserpina during a three day period (that is, during three days and three nights). Moreover, this occurred in the consulship of P. Claudius Pulcher and L. Junius Pullus [249 B.C.] when Rome was suffering from a plague. The Sibylline Books ordered that *stipes* be sent to Father Dis at Terentum. The same books ordered this also, that the children of nobles should sing this hymn on the Capitol."[21]

So writes Pseudo-Acro in commenting on line eight of the hymn composed by Horace for Augustus' celebration of the Saecular Games in 17 B.C. His authority is Verrius Flaccus, the great antiquarian of the Augustan Age, who mentions the Augustan 110-year interval for the *Ludi Saeculares*. But unlike Censorinus, who lays out the fictitious Augustan chronology of the Saecular Games (126, 236, 346, and 456 B.C.), Verrius Flaccus ties his 110-year interval to the year 249 B.C.; and the 110th year before that would have been 362/358 B.C., the year in which, as argued above, the Romans first performed rites to Father Dis and Proserpina at the Tarentum in the Campus Martius. The passage of Pseudo-Acro has its parallel in Censorinus 17.8, which adds a few details from a verbatim quotation of Varro:

> Varro has written thus in the first book of his *De Scaenicis Originibus*: "When many portents occurred, and the wall and tower between the Colline and Esquiline Gates were touched from heaven, and when the *decemviri* therefore approached the Sibylline Books, they reported that the Tarentine Games to Father Dis and Proserpina in the Campus Martius should be performed for three nights, that black victims should be sacrificed, and that the games should be performed every 100 years."[22]

Varro's source of information in this passage is quite specific in mentioning what particular portion of the Servian Wall (including a tower) was struck by lightning. The precision of the information is reminiscent of the more detailed parts of Livy's prodigy lists contained in Books XXI–XLV. We may therefore presume that Varro's information derives directly or indirectly from a documentary source dating to 249 B.C. In its paraphrase of the official pronouncement of the *decemviri sacris faciundis* it uses the term *Ludi Tarentini* to refer to the rites at the Tarentum, and it also indicates that part of this decemviral pronouncement was that the games were henceforth to be performed every 100 years, as if that had not previously been the case.

An examination of the particular events of the First Punic War (and especially of the year 249 B.C.) may help to elucidate the cause and purpose of these rites.[23] At this point the Romans and Carthaginians were in the sixteenth year of a great war, which had required both parties to commit all their considerable resources to the struggle. Despite the ancient and modern stereotype of the land-loving Romans pitted against those highly experienced Carthaginian sea dogs, The Romans had thus far defeated the Carthaginians in every major sea battle of the war. Roman commanders had, of course, experienced reversals in smaller naval encounters, but thus far in the war their huge losses on the sea had resulted from massive shipwrecks caused by storms. Rome's first and only major naval defeat in the war occurred at Drepana in 249 B.C. Drepana had been serving the Carthaginians as an important harbor for receiving supplies by sea in order to maintain the last remaining foothold that the Carthaginians had in the northwestern sector of Sicily. The naval battle there developed when the Roman consul, P. Claudius Pulcher, attempted to carry out a surprise attack upon the Carthaginians. The latter, however, managed to put out to sea in time to engage the Romans; and the end result was that the Romans were decisively defeated. Pulcher escaped with 30 quinqueremes, whereas the other 93 of his fleet were captured by the Carthaginians (Polyb. I.49–51). Since Polybius informs us elsewhere (I.26.7) that the quinqueremes were manned by 300 rowers and 120 marines, the 93 ships captured by the Carthaginians would have involved 39,060 Roman and allied troops. Polybius does not specify how many were killed in the battle, but he says that the crews were captured along with the ships, except for those who had succeeded in running their ships to shore and abandoned their vessels. Even if one-fourth of the rowers and marines had been able to escape capture, it still would have constituted a staggering loss of manpower for the Romans, amounting to nearly 30,000. The defeated consul was soon brought to trial (doubtless by plebeian tribunes before the *comitia centuriata*), and according to Polybius (I.52.2–3) he was forced to pay a huge fine and barely escaped with his life.

As we can judge from the much better documented Hannibalic War of the next generation, the Roman reaction to this naval catastrophe must have manifested itself significantly in the religious sphere.[24] The defeat would have been interpreted as a clear sign that the Romans had somehow lost the *pax deorum*; and all religious resources at hand would have been pressed into service to figure out what had gone wrong, and how the situation could be remedied. The report, mentioned by both Pseudo-Acro and Censorinus in the passages quoted above, of lightning damaging part of the Servian wall and tower must have been viewed symbolically as meaning that the defense of the Roman state was imperiled. Pontiffs, augurs, *decemviri sacris faciundis*, and Etruscan *haruspices* all must have been directed by the senate to investigate the religious state of affairs and to offer advice on what measures needed to be taken in order to restore the *pax deorum* and thereby to ensure Roman victory over Carthage. Given what we know, we may plausibly surmise that Rome's heavy loss in military personnel at Drepana was explained as resulting from the anger of Father Dis and Proserpina, because according to this interpretation, their rites at the Tarentum, which should be performed every 100 years, were ten years overdue. Consequently, the gods of the underworld had displayed their displeasure by snatching away a sizable number of Rome's *iuniores*. Placating Father Dis and Proserpina with lavish ceremonies at the Tarentum during the first three days of June would have been designed to guarantee the safety of Rome's military youth in the future. The seriousness with which the Romans viewed Claudius Pulcher's defeat is seen from a curious incident that occurred three years later. After Pulcher's death, his sister, as she was making her way through a packed crowd at some festival, was overheard to say that she wished that her brother were still alive, because by losing another fleet in Sicily, he could reduce the congested traffic in Rome. The Romans of the day regarded Claudia's wish as no frivolous or laughing matter, because some god unhappy with the Roman state just might bring about its fulfillment. Claudia was therefore prosecuted before the people by aediles and fined 25,000 *asses*.[25]

Another aspect of the *Ludi Saeculares* of 249 B.C. deserves comment. Pseudo-Acro states that on the advice of the *decemviri* a *stips* was to be sent to Dis and Proserpina. This would have been a large collection of small coins contributed by the general public, probably deposited at or near the Tarentum to form a huge pile. We may presume that the collection was lowered into the ground and buried near the Tarentum as a formal offering to the gods of the underworld. According to ancient sources, in the early days of the Republic, when popular leaders such as Publicola or Agrippa Menenius died, the *plebs* made collections of this sort to show their respect and to assist in paying for their funerals. Such collections are also recorded as having been made for offerings to divinities at critical times.[26] Ovid (*Fasti* I.189–193) indicates that it was customary for Romans to give one another a small amount of money, which he terms *stips*, on January 1 as a new year's present for prosperity throughout the coming year. Making such offerings to Father Dis and Proserpina may have been designed to secure health and prosperity throughout the entire *saeculum*. In his explanation for the name of the Lacus Curtius quoted above, Livy says that members of the general public tossed gifts and fruits into the chasm to accompany the self sacrifice of the *iunior*, M. Curtius. According to Suetonius the Roman people every year used to show their respect for the *princeps* in wishing him good health during the year by making offerings of small coins at the Lacus Curtius.[27] This latter practice is likely to have had

Republican antecedents with the Lacus Curtius serving as a kind of wishing well for good fortune; and the similarity between making such offerings at the Tarentum and the Lacus Curtius may have been responsible for Procilius reshaping the observance of *Ludi Tarentini* in 362/358 B.C. into a fanciful story that explained the origin of the name for the Lacus Curtius.

As already discussed, Censorinus in his discussion of the *saeculum* indicates that the Romans adopted the concept from Etruscan lore, and there is no reason to doubt this. The most obvious avenue by which the *saeculum* entered Roman religious thought and practice was the state's consultation of the *haruspices*. Wagenvoort (1956 212–232) has used scenes on a Praenestine *cista* and two Etruscan mirrors, all dating to the fourth century B.C., to develop a most intriguing interpretation, according to which there existed in Etruscan myth a story involving Maris, the child of Minerva, being revived or rejuvenated by being boiled or cooked in a large kettle. The scenes show a baby or young Maris with three different epithets: *Husrnana*, *Halna*, and *Isminthians*. Wagenvoort connects the third with Apollo's Greek epithet Smintheus, the god of plague brought on by mice, well known from the opening lines of the *Iliad*. He equates Etruscan Halna with the Roman *cognomen* Thalna attested among the Juventii, so that Etruscan Halna is equivalent to Roman Juventas. He further elucidates this murky material by citing Aelian's *Varia Historia* IX.16, which ranks Maris among the Ausonians, the oldest of Italy's native inhabitants. The passage says that he lived to be 123 years old, the longest of the Etruscan *saecula* mentioned by Censorinus, and that he died and came back to life three times. Wagenvoort connects Aelian's triple rebirth of Maris with Maris Husrnana, Maris Halna, and Maris Isminthians. He finally concludes that unlike the Romans and other Italic peoples who viewed Mars in part as a spirit of vegetation tied to the annual cycle, the Etruscans associated their Maris with a different temporal cycle, that of the *saeculum*; and that whatever the history of the Roman rites at the Tarentum had been in earlier times, in 249 B.C. they were reorganized, and Maris (or Mars) was replaced by Father Dis and Proserpina. One part of Wagenvoort's thesis involves Mars' association with horses in a chthonic context. In this regard we should recall that in the one etiological story of the Lacus Curtius M. Curtius rides his horse into the chasm and meets his death.

On the other hand, some modern scholars, such as Wuilleumier, Latte, and Orlin (see above n.1), regard the celebration of 249 B.C. as the earliest occasion on which the Romans performed rites to Father Dis and Proserpina in the Campus Martius; and that their designation as *Ludi Tarentini* indicates that they were adopted from the Greek city of Tarentum, because Hades and Persephone are attested as having been recipients of worship in Magna Graecia. Although these arguments are plausible, the fragment of the saecular prayer asking for the continued obedience of the Latins strongly suggests that the Roman rites had an origin prior to 338 B.C. Moreover, since the Tomb of Orcus at Tarquinii, dating sometime from the middle of the fifth to the end of the fourth century B.C., depicts an Etruscan wolf-headed Hades and his divine consort presiding over the underworld, the Roman rites to Father Dis and Proserpina need not have been a mid-third century importation from the Greeks of southern Italy.

THE EXPIATION OF 207 B.C.[28]

"1. Before the consuls set out, there was a nine-day ritual because it had rained stones from heaven at Veii. 2. As happens, as the result of the mention of the single prodigy, others were also reported. Lightning struck the temple of Jupiter and the grove of Marica at Minturnae, as well as the city wall and a gate at Atella. 3. The people of Minturnae added what was even more dreadful, that a stream of blood had flowed on the gate. At Capua a wolf by night had entered a city gate and had mutilated a guard. 4. These prodigies were expiated with full grown victims, and there was a *supplicatio* of one day in accordance with a decree of the pontiffs. Then the nine-day ritual was repeated because it seemed to rain stones in the Armilustrum. 5. After people's minds had been freed from religious scruple, they were again disturbed by a report that at Frusino there had been born an infant the size of a four-year old. It was not so much his size which caused amazement, but the fact that he had been born of indeterminate sex, just like one two years before at Sinuessa. 6. Accordingly *haruspices* were summoned from Etruria and pronounced it to be a foul and awful prodigy. It had to be removed from Roman territory, far from contact with the land, by being drowned in the sea. They placed it living in a chest and cast it out to sea. 7. The pontiffs also decreed that thrice nine maidens should go throughout the city singing a hymn. While they were in the temple of Jupiter Stator learning the hymn composed by the poet Livius, the temple of Juno Regina on the Aventine was struck by lightning. 8. After the *haruspices* had replied that this prodigy was related to mothers, and that the goddess had to be placated with a gift, 9. a decree of the curule aediles summoned to the Capitol the women who lived in the city of Rome or within the tenth milestone. They chose 25 from among themselves to whom they were to bring donations (*stipem*) from their dowries. 10. From this a gift of a golden pelvis was made and brought into the Aventine shrine, and a sacrifice was properly performed by the matrons. 11. The *decemviri* immediately decreed a day for another sacrifice to the same goddess. Its procedure was as follows. Two white cows were led into the city from the temple of Apollo by way of the Porta Carmentalis. 12. Two statues of Queen Juno made of cypress wood were carried behind them. 13. Then came the 27 maidens dressed in long gowns, singing their hymn to Queen Juno. For that time the hymn was perhaps praise worthy to uncultivated minds, but it would now be rough and uncouth if recorded. The *decemviri*, wearing laurel crowns and clad in their *toga praetexta*, followed after the line of maidens. 14. From the gate they came into the Forum by way of the Vicus Jugarius. The procession halted in the Forum, and a rope was passed through their hands. The maidens advanced, keeping their singing in time with the beating of their feet. 15. From there, by way of the Vicus Tuscus and the Velabrum, they came through the Forum Boarium and into the shrine of Queen Juno along the Clivus Publicius. There the *decemviri* offered up the two sacrificial victims, and the two statues of cypress wood were carried into the temple."[29]

Livy's account of these ceremonies is unusually detailed. Whatever his immediate source of information happened to be, it apparently even contained the text of Livius Andronicus' hymn, but unfortunately for us, Livy chose not to include it so as not to bore his contemporary Roman readers. The passage illustrates the workings of the official Roman religious machinery, involving the expert advice and participation of Etruscan *haruspices*, pontiffs, and the *decemviri sacris faciundis*. When the *haruspices* were consulted concerning the birth of a hermaphrodite, they declared it to be an abomination that must be removed entirely from Roman territory by being drowned in the sea. Meanwhile, the pontiffs decreed that the city (and hence the Roman state) must be cleansed of the pollution through a ceremony involving thrice nine maidens; but as the latter were practicing their hymn in the temple of Jupiter Stator, the shrine of Juno Regina on the Aventine was struck by lightning. Whereupon the *haruspices* were again consulted; and when they advised that Juno Regina must be placated with a gift, the curule aediles cooperated with the matrons of Rome to have a suitable offering prepared. What eventually emerged was that the ceremony of thrice nine maidens originally decreed by the pontiffs was augmented by another one resulting from the second consultation of the *haruspices*; and the *decemviri* organized, supervised, and participated in the entire observance.

The *haruspices* clearly viewed lightning striking Juno Regina's temple as evidence that the proper ritualistic disposal of the hermaphrodite had not gone far enough to restore the breach in the *pax deorum*, and that additional measures needed to be taken. The offering of a golden pelvis was designed to persuade Juno to bless Roman women with normal childbirths and is entirely consistent with archaeological finds at ancient shrines of countless offerings of clay or metal objects shaped like body parts. The religious procession began at the temple of Apollo Medicus in the southern part of the Campus Martius near the Capitoline in order to enlist the aid of Apollo as a god of healing. The use of thrice nine maidens to assist in purifying the state is reminiscent of the thrice nine *Argei* employed by the Romans for the great purification of mid May, discussed in the preceding chapter. Finally, since the cypress tree was associated with the dead, the two statues of cypress wood offered to Juno must have atoned for the two hermaphrodites, the recent one born at Frusino and the other one born two years before at Sinuessa.

What makes this ceremony, however, relevant to the Saecular Games is the procession of thrice nine maidens and their singing of a hymn composed for the occasion by Livius Andronicus, a leading poet of the day. The Augustan *Acta* of the *Ludi Saeculares* (*ILS* 5050 ll.147–149) indicate that on the third and final day, following sacrifices and prayers to Apollo and Diana at Augustus' new temple to Apollo on the Palatine, thrice nine boys and thrice nine maidens sang the *Carmen Saeculare* composed by Horace. It therefore looks as if the Augustan Saecular Games included some aspects of the religious ceremony devised to purify Rome in 207 B.C. Unfortunately, we have no way of knowing whether this was an Augustan innovation or was taken over from the Saecular ceremonies observed in 146 B.C. The following quotation from Livy (XXXI.12.6–10) for the year 200 B.C. shows that the expiatory observance of 207 B.C. had been something of an innovation itself and had formed an important precedent for later times:

> 6. Next there were reported from many places the obscene births of animals. Among the Sabines was born an infant, it being uncertain whether it was male or female. Another one of ambiguous sex had also been discovered sixteen years earlier. 7. At Frusino there was born a lamb with a pig's head; at Sinuessa a pig born with a human head; on public land in Lucania a colt born with five feet. 8. All were viewed as foul, deformed, and of nature going astray in producing hybrid offspring. Above all the hermaphrodites were declared to be abominations and were ordered to be carried out to sea forthwith, just as a similar monstrous birth had been deported recently in the consulship of C. Claudius and M. Livius [207 B.C.]. 9. Nevertheless, they ordered the *decemviri* to approach the books concerning the portent. According to the books the *decemviri* ordered the same rites that had been performed recently following that prodigy. In addition, they ordered a hymn to be sung by thrice nine maidens throughout the city, and that a gift be carried to Juno Regina. 10. In accordance with the reply of the *decemviri* the consul C. Aurelius saw that these things were done. Just as Livius had done within the memory of the senators, so now P. Licinius Tegula composed the hymn.[30]

Thanks to brief notices in Julius Obsequens' Book of Prodigies excerpted from Livy, we know that the procession of thrice nine maidens through Rome was performed on at least five other occasions during the late Republic: in 133 following the murder of Ti. Gracchus and many of his supporters, and following the birth of a hermaphrodite at Ferentinum and it being cast into the river (27A), in 120 following an even larger massacre of Romans with C. Gracchus, and following the discovery and casting into the sea of an eight-year old hermaphrodite (34), in 117 following the death of the consul M. Porcius Cato in battle against the Scordisci, and following the discovery and drowning in the sea of a ten-year old hermaphrodite found at Saturnia (36), in 104 following the Roman disastrous defeat at Arausio (43), and in 99 following another massacre of Romans in the city resulting from the sedition of Saturninus and further civil discord over an agrarian law proposed by the tribune Sex. Titius (46). Obsequens records numerous monstrous births throughout his brief catalogue of prodigies, but only in the instances just listed does he record that they were cast out to sea and were accompanied by a purification of the city by thrice nine maidens singing a hymn. Moreover, under the year 104 B.C. he records the additional advice of the *haruspices* for the people to offer a *stips* to Ceres and proserpina, obviously intended to placate the deities of the underworld for the great losses at Arausio in the previous year. Similarly, in 99 the people again offered a *stips* to Ceres and Proserpina in accordance with the advice of the *Haruspices*, and images of cypress wood were given to Juno Regina. From these instances it appears that during the late Republic the procession of trice nine maidens was not simply used to cleanse the state of the pollution of a hermaphrodite, but more importantly, it was employed to purify the state from the bloodshed of civil violence in Rome, as well as to assist in the restoration of the *pax deorum* after a major military defeat. Indeed, it seems likely that in the wake of civil bloodshed or military disaster dire warnings, such as hermaphrodites, were discovered *post eventum* in order to account for the calamity that had befallen the Roman state. Thus, Augustus might have incorporated the ceremony of thrice nine maidens singing an expiatory hymn into his celebration of the Saecular Games in order to rid the state of the pollution of civil war and bloodshed and to inaugurate his new era of stability and peace.

Finally, it should be noted that the expiation of 207 B.C. probably occurred in the year in which the sixth Etruscan *saeculum* came to an end. Censorinus (17.5–6) says that since humans are often unaware when a *saeculum* ends, the gods send portents as reminders; and that the Etruscans observed and recorded such things. We may therefore wonder whether the *haruspices* consulted by the Romans in 207 B.C. regarded the monstrous birth in this context, and whether Roman knowledge of its possible connection with the ending of an Etruscan *saeculum* contributed to the singing of a hymn by thrice nine maidens being absorbed into the Roman *Ludi Saeculares*.

THE SAECULAR GAMES OF THE LATE REPUBLIC

The Ludi Saeculares were due to be repeated in 149 B.C. Censorinus (17.11) says that according to Varro, Valerius Antias, and Livy they were in fact observed that year; but then he cites L. Calpurnius Piso, Cassius Hemina, and Cn. Gellius (the former two at least being adults at this time) for the games' celebration three years later in 146. It would be remarkable if the great Varro was in error on this point. Perhaps Censorinus or his sources cited for 149 B.C. misconstrued a statement that Rome's *saeculum* ended that year into a brief notice to the effect that the *Ludi Saeculares* were then celebrated. In any event, the games were postponed for three years, doubtless due to the Roman state being fully occupied with the wars in Greece and North Africa. Although we might suppose that the Romans would have wished to observe the Saecular Games on schedule so as not to risk endangering the *pax deorum* in serious times of war, there might have been very practical considerations behind the delay, because many of Rome's priests necessary for organizing and administering the celebration might not have been present in the city due to service abroad in the armies as elected magistrates or appointed legates. It should also be remembered that Rome was fighting as well on a third front in both Spanish provinces. By 146 B.C., however, the two wars in Greece and North Africa had been concluded decisively in Rome's favor with the destruction of both Carthage and Corinth, a most striking demonstration of Rome's mastery throughout the entire Mediterranean. In addition, the victories resulted in the creation of two new provinces: Macedonia and Africa. Consequently, although we have no source of information concerning the nature of the *Ludi Saeculares* observed in 146 B.C., we may surmise that if the celebration took place after the news of Rome's victories in Greece and North Africa, they might have differed markedly from the Saecular Games of 249 B.C. in that they were performed in a climate of exhilaration and were seen as ushering in a new age of Roman world dominion.[31] If so, the climate would have lent itself to innovations being made in the ceremonies to accommodate and express these ideas.

Rome's next *saeculum* was due to end in 49 or 46 B.C. A series of coins, dated by Crawford to 45 B.C. and issued by L. Valerius Asciculus, clearly allude to the ending of a *saeculum* and anticipating the celebration of the Saecular Games (Crawford 1974 483–485). One type is engraved with the head of a Sibyl, others with the head of Apollo, the god of prophecy and hence of the Sibylline Books. Another type bears a cornucopia, probably referring to the supposed prosperity of the new *saeculum*; and another type shows Sol on one side and Luna on the other, symbolizing Apollo and Diana, who figured prominently in the Augustan Saecular Games. Servius, commenting on Vergil's Ninth Eclogue l.46, records a curious story. When Octavian was holding funeral games for Julius Caesar in 44, a star suddenly appeared in the sky and was visible during the daytime. Octavian regarded it as Caesar's soul having ascended into heaven, so that a golden star was then placed above the head of Caesar's statue on the Capitoline. A *haruspec*, however, named Vulcanius, declared in a public meeting that it was a comet that marked the end of the ninth and the beginning of the tenth *saeculum*; and since he was revealing secrets against the will of the gods, he said that he would die immediately. As soon as he had finished speaking, he fell dead in the assembly.[32]

Despite the ending of a *saeculum*, the turmoil resulting from the civil war between Caesar and Pompey, followed by further chaos after Caesar's assassination, caused the games not to be celebrated at any point during the decade of the 40's B.C. Nevertheless, Vergil's Fourth Eclogue, written to honor Asinius Pollio's assumption of the consulship in 40, expresses the hope of a new *saeculum* that would abolish the miseries of the triumviral period and would bring back the golden age of the distant past. The poem clearly reflects the Romans' hope for the return of peace with the treaty of Brundisium and the marriage of Antony and Octavia (Duquesnay 1977 32–34); but the poem's notion of a new golden age may exhibit the influence of Hellenistic religious ideas, especially from Egypt, where the accession of a new monarch to the throne had long been celebrated as marking the advent of a new age of prosperity.[33] This new view of the *saeculum* coincided with the death of the Republic and the rise of dynasts who were attempting to establish themselves as monarchs along the lines of Hellenistic kings. With the final defeat of the Republican cause at Philippi in 42 Antony and Octavian, drawing upon the religious and artistic traditions of the Hellenistic East, increasingly had themselves represented as god-like saviors coming to rescue a war-weary world (Zanker 1988 33–77).

THE AUGUSTAN AND SEVERAN SAECULAR GAMES

As in so many other facets of Roman public life, Augustus' celebration of the *Ludi Saeculares* formed an important precedent that was followed by later emperors. The games were probably also typical of Augustus' handiwork in that they were a complex blending of old hallowed traditions suffused with innovations in order to preserve the former, to maintain continuity with the past, and to make the institution relevant and applicable to the new political reality of the principate. This pattern in the area of religion is most clearly seen in Augustus' reorganization of the cult of the *Lares Compitales* and of the priesthood of the *Fratres Arvales*. In the former case Augustus incorporated into his systematic restructuring of the city of Rome, henceforth and even to this day divided into fourteen administrative regions, people's traditional worship of the Lares of their own neighborhood (*vicus*) at small shrines set up at the intersections of streets (*compita*). The chapels were rededicated to the *Lares Augusti*, and the *magistri* of the *vici* henceforth made offerings for the emperor's health at these shrines to the *genius* of Augustus.[34] The reorganization thereby combined the traditional worship of the *Lare Compitales* with a new mechanism for expressing and reinforcing loyalty to the *princeps*.

Thanks to Augustus' revival of many obsolescent aspects of the Roman state religion, the priesthood of the *Fratres Arvales*, attested in only one ancient passage dating to the Republic (Varro *L. L.* V.85), is well known to us from a corpus of inscriptions found at the ancient site of the sacred grove of the obscure agrarian divinity Dea Dia, whom the *Fratres Arvales* worshipped, and to whom they performed rites for the fertility of the crops. The earliest of these inscriptions dates to c.20 B.C., and the latest to c.300 A.D. The inscribed *Acta* of these priests' religious activities present us with a curious mixture of antiquated agricultural rites and sacrifices and prayers offered for the welfare of the emperor and members of his family.[35] We may therefore plausibly suppose that in observing the Saecular Games Augustus probably adapted the ceremonies in some respects in order to express the ideology of the new political order that he was establishing.

The Augustan principate, of course, did not spring into existence fully formed overnight.[36] Four months after the battle of Actium (fought on

September 2, 31 B.C.) Octavian entered upon his fourth consulship, during which he pursued the defeated Antony and Cleopatra to Egypt, where the latter two eventually committed suicide. Octavian then annexed Egypt as a province and returned to Rome. Like Caesar following the civil war with Pompey, Octavian found himself in full mastery of the Roman Empire, but was faced with the dilemma of what to do next, and how to consolidate and regularize his autocratic position. At first he simply had himself reelected as one of the two annual consuls, which *ex officio* made him one of the two heads of state, but in 28–27 B.C. Octavian officially renounced his extra-legal position as *dux* that he had assumed (or usurped) on the eve of the Actium campaign (see *Res Gestae* 25). As Augustus himself later wrote (*Res Gestae* 34.1): "In my sixth and seventh consulships, after I had put an end to the civil wars, having obtained all affairs through the consent of everyone, I transferred the *res publica* from my own power into the judgment of the senate and Roman people = In consulatu .sexto et septimo, postquam bella civilia exstinxeram, per consensum universorum potitus rerum omnium, rem publicam ex mea potestate in senatus populique Romani arbitrium transtuli." Following this restoration of constitutional government, the senate bestowed the honorific name of Augustus upon Octavian and also granted him an extraordinary proconsular command for ten years to govern the provinces of Spain, Gaul, and Syria, which during the pinnacle of the first triumvirate's power had been governed by Pompey, Caesar, and Crassus respectively. The proconsular command legalized Augustus' control of virtually all Roman military forces, and he continued to be elected consul until 23 B.C., at which time aristocratic dissatisfaction with his dominance of the consulship convinced Augustus that he needed to be less blatant in heading the Roman state. He therefore abandoned the consulship as a key ingredient of his legal position, but he was given the power of a plebeian tribune, which he then held for the rest of his life, and by which he henceforth numbered the years of his principate. Then in 19 B.C., Augustus' legal position was further enhanced when he received the privileges and powers of a consul, which, like the tribunician power, was separate from the office itself, and which he held for the rest of his life. Thus, by the year 19 the three principal powers upon which Augustus legally founded his position as *princeps* (the powers of a consul and plebeian tribune along with an extraordinary proconsular command renewed at intervals of five or ten years) were set in place. Accordingly, as Jones has argued, we can date the inception of the Augustan principate in terms of its legal basis to 19 B.C.

During the following year (18 B.C.), Augustus further elaborated his principate by clearly designating Agrippa to be his partner and successor: for like Augustus, he was granted the power of a plebeian tribune and was also given an important proconsular command. Then after conducting a *lectio senatus*, Augustus passed two laws: the *Lex Iulia De Adulteriis* and the *Lex Iulia De Ordinibus Maritandis*.[37] They formed the cornerstone of Augustus' attempt to reverse negative demographic trends that had become widespread in Roman upper class society during the late Republic: divorce, bachelorhood, sexual promiscuity, and a low birth rate. Augustus doubtless believed that the effective continuance of the Roman Empire required the Roman upper class, whose members formed the ruling elite, to maintain a healthy reproductive rate.

Also under the year 18 B.C. Dio (LIV.17.2, cf. Suet. *Aug.* 31.1) records that Augustus ordered the *quindecimviri sacris faciundis* to produce an authoritative text of the Sibylline Books, which henceforth were to be under their exclusive control. This measure was obviously taken with a view to the celebration of the *Ludi Saeculares* in the next year. Moretti's restored text of both old and new fragments of the Augustan *Acta* (1985 366) contains a decree of the senate dating to this year that concerned itself with making the necessary financial arrangements for the *Ludi Saeculares* scheduled for the next year. The resolution indicates that a search through the records of the *quindecimviri sacris faciundis* had failed to turn up any information as to how the preceding Saecular Games had been funded. The senate therefore was left to its own devices in resolving the matter. In addition, Zosimus (II.6) records the text of the Sibylline Books pertaining to these games; and since he mentions the Augustan jurist Ateius Capito, and since the oracular verses regard a *saeculum* as lasting 110 years, it is evident that this Sibylline text was reworked at this time. Our suspicions are confirmed by a comparison of these Sibylline verses with the epigraphic text of the Augustan *Acta*, because the two correspond to one another in many details, such as having the matrons kneel while uttering their prayer to Juno Regina.

After consolidating and regularizing his constitutional position in 19 and ensuring the reproductive viability of the Roman upper class through the passage of his laws in 18, Augustus proceeded in the next year to orchestrate his celebration of the *Ludi Saeculares*. Given the latter's association with the promotion of childbearing and protection against epidemic diseases, the Saecular Games would have been ideal in underscoring Augustus' social legislation. Moreover, the idea of a *saeculum* commencing a new age of peace, prosperity, and moral goodness, as outlined in Vergil's Fourth Eclogue, would have also been consonant with Augustus' desire to use a hallowed Roman religious tradition to cleanse the state of the pollution of civil bloodshed and to announce the beginning of a new age of stability and tranquility. But in order to justify celebrating the Saecular Games at this time Augustus was compelled to concoct a fictitious chronology of the games based upon a *saeculum* of 110 years. Taylor (1934 119) pointed out that two of the *quindecimviri sacris faciundis* at this time were patrician Valerii, who were probably eager to promote their family's historical association with the games (already well pioneered by Valerius Antias). Is it therefore coincidental that according to the Augustan chronology the first and second celebrations occurred in 456 and 346 B.C. when a Valerius was consul? Since Censorinus cites the *commentarii* of the *quindecimviri* for the dates of the Augustan fictitious chronology, it is evident that it became an integral part of their historical records; and these dates were given further official endorsement by being inscribed in their own section of stone on the *Fasti Consulares* that Augustus had erected in the Forum.

If Augustus and his fellow priests did not scruple to manipulate the chronology of the earlier celebrations, how else might they have changed the nature or rites of the games? The Augustan Acta record the following sequence of events.[38] Augustus commenced the celebration in the evening of May 31 in the Campus Martius by praying and sacrificing nine female lambs and nine female goats to the Moerae, the goddesses of fate and hence of the Sibylline Books preserved and used by the *quindecimviri* for the welfare of the Roman state. The sacrifices were followed by stage performances and then a *sellisternium* to Juno and Diana conducted by 110 chosen Roman matrons. June 1 began with Augustus and Agrippa each sacrificing an ox to Jupiter Optimus Maximus on the Capitoline. This was followed by stage performances in a wooden theater constructed in the Campus Martius and another *sellisternium* held by the matrons. The games went on into the night, during which Augustus prayed and offered up to the ilithyae three different kinds of cakes, nine of each sort. June 2 began with Augustus and Agrippa each sacrificing a cow on the Capitoline to Juno Regina. Their prayer was followed by another spoken by the 110 matrons. Then as games were again being performed in the Campus Martius by night, Augustus prayed and sacrificed a pregnant sow to Terra Mater, followed by another *sellisternium* conducted by the matrons. June 3 began with Augustus and Agrippa praying and offering to Apollo and Diana on the Palatine the same 27 cakes that had been previously sacrificed to the Ilithyae. Next the thrice nine boys and thrice nine maidens sang the *Carmen Saeculare* composed by Horace. The stage performances that night in the Campus Martius were followed by chariot races and other equestrian events involving *desultores*. The Augustan *Acta* then end with two decrees of the *quindecimviri*, announcing the addition of several more days of festivities for amusing the public.

The single most extraordinary feature of this synopsis is the total absence of Father Dis and Proserpina, who, according to the ancient sources, were the only two divinities associated with the *Ludi Saeculares* during Republican times. Secondly, Jupiter Optimus Maximus and Apollo are the only two male divinities mentioned in these rites. All others are female and are associated with fertility and childbirth, which fits perfectly with Augustus' social legislation of the previous year. On the other hand, they are also consistent with the overlap or confusion in the ancient sources of the earliest Saecular Games and the *Ludi Taurei*. The presence of Jupiter Optimus Maximus, receiving sacrifice on the first day of the three-day celebration, is not surprising, because he was after all the chief tutelary deity of the Roman state. Apollo's presence can be accounted for in three ways. First of all, his role as the god of prophecy naturally associated him with the Sibylline Books. Secondly, as god of the sun, whose motions mark the passage of time, he could have been easily identified with the concept of the *saeculum*. In fact, as noted above, one coin type of P. Valerius Asciculus, dating to the time of Caesar's dictatorship and probably alluding to the ending of the *saeculum* at that time, portrays the sun on one side and the moon on the other, suggesting that even before the Augustan celebration in 17 B.C. both Apollo and Diana were connected with the Saecular Games. Thirdly, given Apollo's importance as a god of healing, and what the ancient sources tell us about the origins of the Saecular Games, his presence in the rites is quite understandable; but this is not Apollo Medicus of the southern Campus Martius, whose temple had been the starting point for the expiatory procession in 207 B.C., but this is Palatine Apollo, the god of Augustus' new splendid temple vowed at Actium, the victory upon which the Augustan principate was founded.

Tellus, not Terra Mater, was traditional in Rome's state religion. She and Ceres were worshipped together in late January during the *Feriae Sementivae*, the *Sowing Festival*. Like Terra Mater in the Augustan Saecular Games, she received the sacrifice of a pregnant sow to ensure the fructifying powers of earth with the advent of a new spring (Ovid *Fasti* I.657–704). Terra Mater, therefore, appears to be a later intrusion. Unfortunately, in the present state of our evidence, including our complete ignorance concerning the rites conducted in 146 B.C., we are at a loss to determine precisely what else might have been added by Augustus and his priests. It is likely that the games were reinterpreted and modified in some degree each time that they were celebrated in order to conform to existing views and needs. In any case, the Augustan celebration does not appear to have possessed any gloomy features as the supposed origins of the games would have suggested; and we must even entertain the possibility that the role of Father Dis and Proserpina is a later Roman invention designed to give the games a somewhat sinister and exciting origin. In fact, Valerius Antias would be a good candidate for such a tale, because one of his fragments (Arnobius V.1) shows that he recorded in his history a similar spooky encounter between King Numa Pompilius and Jupiter Elicius.

One interesting aspect of the Imperial Saecular Games concerns the offering of first fruits by the general citizenry and the distribution of purifying agents by the *quindecimviri*. According to Zosimus (II.5) some days before the actual celebration the *quindecimviri*, seated on the podia of temples on the Capitoline and palatine, distributed to the general public purifying agents: torches, sulphur, and bitumen. These details are confirmed by the surviving portion of the Severan *Acta* (*ILS* 5050A), which describe how the *quindecimviri* assembled in the Palatine temple of Apollo on May 25 to cast lots to determine who would sit on which tribunals to distribute *suffimenta*. We find similar provisions in the Augustan Acta (*CIL* VI. 32,323 ll.30–50), which specify the giving out of purifying agents (*purgamenta dari*) and the acceptance of first fruits (*fruges accipere*), immediately after which comes mention of some of the stations at which these procedures were to take place: in front of the Capitoline temple of Jupiter Optimus Maximus, at the shrine of Jupiter Tonans (also located on the Capitoline), at Apollo's Palatine temple and its portico, and at Diana's shrine on the Aventine and its portico. Following these provisions come others to be performed over several days before June 1, particularly the distribution of *suffim[enta]*, after which comes mention of the Saepta, obviously one of the places designated to serve as a station for distribution.

The handing out of *suffimenta* to the general populace from tribunals is depicted on coins commemorating the Saecular Games. One Augustan coin type shows the emperor seated upon a tribunal and handing out materials to two people standing before him. The tribunal is labeled "Lud(is) S(aecularibus)," thus clearly indicating that the coin and the ceremony shown refer to the Saecular Games. The coin's legend reads: "Aug. suf. p. = Aug(ustus) suf(fimenta) p(opulo distribuit)." Several coin types of Domitian portray scenes from his celebration of the Ludi Saeculares in 88 A.D. One closely resembles that of Augustus just mentioned and bears the legend: "Cos. XIII Lud(is) Saec(ularibus) suf(fimenta) p(opulo) d(istribuit) s(enatus) c(onsulto) = in his fourteenth consulship, by decree of the senate, during the Saecular Games he distributed *suffimenta* to the people."[39] *Suffimentum* is a relatively rare word, and in Imperial Latin is a general term for a purifying agent (*purgamentum*). The word derives from *sub-* and *fumus* and should specify things that burn and create smoke, which is in accord with Zosimus' list of torches, sulphur, and bitumen.

Ovid's account of the Parilia of April 21 is the only surviving ancient Latin passage containing the even rarer word *suffimen* (*Fasti* IV.731–734). The rites on this day were originally intended to purify the farmer's herd animals before letting them out into summer pastures. The ceremony consisted of driving the animals over a fire of bean straw (Ovid *Fasti* IV.721–782), but the rites performed by the Roman priests on this day included another item, what Ovid calls *suffimen*, the dried blood from the sacrificed October Horse of the previous October 15 mixed with the ashes of the Fordicidia of April 15, observed six days before the Parilia and exactly six months before the rites of the October horse. On the Fordicidia pregnant cows (*fordae*) were sacrificed to ensure the earth's fertility, and the unborn calves were removed and burned (Ovid *Fasti* IV.629–640). On the Parilia of April 21 the *Vestal Virgins* gave out to people minute amounts of this precious *suffimen* that was to be thrown in the fire of bean straw, apparently in order to increase its purificatory efficacy.

From the very beginning of Roman historical writing, the Parilia of April 21 was regarded as the day on which Romulus founded the city of Rome (Plut. *Rom.* 14.1, citing Fabius Pictor). Consequently, this agrarian ceremony of purifying farm animals was transformed into an urban celebration of Rome's birthday. We do not know when *suffimenta* became incorporated into the *Ludi Saeculares*, but it could have been an Augustan innovation. Thanks to Ovid's account of the Parilia, *suffimen* and *suffimenta* tie together Rome's birth and the *saeculum* and thus leads us naturally to think in terms of birth and rebirth. The idea is clearly present in the

Severan *Acta*. Just before the epigraphic text breaks off, it records the locations of the first two tribunals assigned by lot and from which *suffimenta* were to be distributed. The first such tribunal was located in the area of Palatine Apollo, whereas the second one was also on the Palatine at Roma Quadrata, the supposed original site of Romulus' settlement (Festus 312L s.v. *Quadrata Roma*, quoting Ennius). Since Augustus was most certainly responsible for including Palatine Apollo in the *Ludi Saeculares*, he might have also been the first to associate the *saeculum* with Roma Quadrata and the idea of Rome's rebirth. This definitely was in accord with Augustus' own self image as Rome's new founder. According to Dio (LIII.16.7), when in 27 B.C. Octavian was considering what new name he should adopt to mark the beginning of his new regime, he seriously thought of taking the name Romulus, but he eventually decided upon Augustus. Suetonius (*Aug.* 7) says that the latter was the suggestion of Munatius Plancus, who supported his suggestion by citing Ennius' account of Romulus founding Rome "augusto augurio."

The notion that the Imperial *Ludi Saeculares* embodied Rome's birth and rebirth is evident from another parallel between Rome's foundation story and the rites of the Imperial Saecular Games. According to Plutarch (*Rom.* 11) and Ovid (*Fasti* IV.819ff), when Romulus marked out the site for his new community, he did two things. First, he dug a trench into which his followers threw first fruits and portions of soil from neighboring areas, after which the trench was refilled. Next, he plowed the so-called first-born furrow of the *pomerium* (*sulcus primigenius*) and constructed Rome's first defensive wall. Plutarch further adds that the initial trench was dug in the Forum, was circular in shape, was called the *Mundus*, and corresponded to the Comitium. Tossing first fruits into the trench resembles Livy's account of the Lacus Curtius in 362/358 B.C. and also reminds us of the *stips* offered to Father Dis and Proserpina mentioned by Pseudo-Acro in connection with the rites observed in 249 B.C. In his description of the Imperial ceremonies of the Saecular Games Zosimus (II.5) says that people were urged to bring from home wheat, barley, and beans, which they were to offer as first fruits either at Diana's Aventine temple or at the stations where the *quindecimviri* were distributing *suffimenta*. As already mentioned, the Augustan *Acta* specify that while at their assigned stations for distributing purifying agents, the *quindecimviri* were to receive *fruges*. One of Domitian's Saecular coin types portrays this procedure: as the emperor sits in a chair on a low platform with a temple in the background, two people standing before him offer their first fruits by pouring them onto a heap on the ground (Mattingly-Sydenham 1926 201 #375). The coin's legend reads: "Cos. XIIII Lud(is) Saec(ularibus) a p(opulo) f(ruges) ac(ceptae) s(enatus) c(onsulto) = In his fourteenth consulship, by decree of the senate, during the Saecular Games first fruits were accepted from the people." It therefore looks as if the Imperial *acceptio frugum* was an adaptation of placing such offerings in the *Mundus* or the *sulcus primigenius* of the *pomerium* when

Rome was founded. The innovation enabled a large number of people to participate directly in a ceremony that through the imitation of Rome's birth assured Rome's rebirth and security for yet another *saeculum*.

Finally, as other modern scholars of the Saecular Games have pointed out, the notion of the Augustan *Ludi Saeculares* marking Rome's rebirth is suggested by Augustine's citation of Varro's *De Gente Populi Romani* in XXII.28 of the *De Civitate Dei*. In this passage, the Christian polemicist cites the great antiquarian concerning a Greek belief of rebirth, according to which 440 years after people's deaths their souls are reunited with their bodies to live again. It certainly cannot be an accident that Augustus' fictitious chronology of the Ludi Saeculares extended back into Rome's past exactly 440 years (456–17 B.C.).

EPILOGUE

Toward the end of his discussion of the *saeculum,* Censorinus (17.15) cites Varro for the following notion. Vettius, an acquaintance of Varro's and a man very skilled in augury, said that if Romulus had in fact seen twelve vultures when he and Remus were divining which of them should be Rome's first king, then Rome would endure for twelve *saecula*. Calculating from c.750 B.C., we come to c.450 A.D. for Vettius' prediction of Rome's termination. It turned out to be fairly accurate. Alaric and the Visigoths sacked Rome in 410, which marked the beginning of the end. By 450 the western provinces had been overrun by different Germanic tribes, and Roman Imperial rule throughout the region had vanished. In 455, Rome was sacked again, this time by Gaiseric and the Vandals from their kingdom in North Africa; and in 476 Romulus Augustulus was forced by another Germanic king, Odovacer, to abdicate as Rome's last emperor of the West. Zosimus, an unreformed pagan, who probably wrote his history during the second half of the fifth century, attributed the demise of the Roman Empire to Diocletian's and Constantine's failure to celebrate the *Ludi Saeculares* when they should have been repeated c.304–313 (using both 100 and 110-year intervals): for as he observed, as long as the Romans maintained these and other traditional practices, the empire was preserved.

5 Magna Mater and the Taurobolium[1]

On the eve of the battle at the Mulvian Bridge, fought just north of Rome in October of 312 A.D., Constantine imitated countless Roman generals before him in seeking divine assistance to achieve victory over his adversary. Yet, unlike other earlier vows made by commanders in the field, Constantine's recourse to the god of the Christians did not simply result in the erection of a temple in Rome, whose day of dedication was henceforth commemorated as the cult's *dies natalis* in the Roman religious calendar. Indeed, Constantine's choice of divinities had far more lasting consequences. For the past 250 years, Christians had suffered from official persecution, largely intermittent and occurring at the local level, often arising from specific local incidents or crises. The near collapse, however, of the Roman empire during the mid-third century A.D. had produced the first systematic persecution by the Roman state, stimulated by a widespread popular belief that the military and political turmoil throughout the empire was due to the wrath of the gods, brought on by the Christians' refusal to accord them their proper worship. A second great persecution had been set in motion more recently by the Emperor Diocletian, commencing on February 23 of 303 and officially halted by Galerian in 311.[2]

Constantine's victory over his imperial[3] rival Maxentius at the Mulvian Bridge secured his power over the western half of the empire, and it was soon followed by an official pronouncement, jointly agreed upon by Constantine and his imperial colleague of the East, Licinius, during a meeting in Milan in early 313. This so-called Edict of Milan ordained that property recently confiscated from Christians should be restored, and henceforth Christians would be allowed to practice their religion without hindrance.[4] Over the course of the fourth century, however, Christianity rapidly progressed from a religion tolerated by the Roman state to the only religion permitted by the Roman state. This process officially culminated in the famous decree of the Emperor Theodosius, promulgated on February 24 of 391 (*Codex Theod.* XVI.10.10), which categorically forbade the performance of pagan sacrifices and the use of pagan shrines for worship. Thus, the struggle between Christianity and paganism during the fourth century A.D. forms one of the most important chapters in the intellectual history of Western Civilization.

Christianity's ultimate success and triumph over paganism was largely, if not entirely, due to its monotheistic intolerance of other forms of religion, a characteristic inherited from Judaism: "Thou shalt have no other gods before me" (*Exodus* 20.3). A blissful existence after death was a message not the least bit monopolized by Christianity, since it had long been a promise held out by several pagan mystery religions. In general, the latter had begun as localized cults that had spread beyond their parochial borders during Hellenistic or Roman Imperial times. The spread and interaction of different religious traditions with one another created a rich and complex amalgamation of ideas. In some instances local cults were transformed and universalized by adopting features compatible with the commonly shared Hellenistic religious culture.[5] In fact, Christianity can be viewed as a kind of Jewish mystery religion, the product of Judaism cross-fertilized with Hellenistic religious ideas. At least three pagan religions are known to have promised a happy afterlife to their initiates long before the advent of the Hellenistic age. The Pyramid Texts of Egypt demonstrate that Osiris was a god of personal salvation as early as the middle of the third millennium B.C.; and both Orphism and the Eleusinian Mysteries flourished in the Greek world during the classical period. These religions, however, are rather exceptional in that a sufficient body of ancient evidence allows us to reconstruct in broad outline their theology and religious history. Unfortunately, in other instances, such as Mithraism and the cult of Magna Mater, secrecy proved all too successful in shrouding their religious doctrines in mystery with the result that the existing ancient data leave many fundamental questions of their theology and religious history unanswered.

PRIMITIVE ORIGIN OF THE TAUROBOLIUM

While foundations were being dug for the facade of the Basilica of St. Peter in the Vatican in 1609, workmen unearthed a series of eight inscribed marble altars commemorating the initiation of Roman senators into the rites of Magna Mater through the performance of a taurobolium and/or criobolium (i.e., a special kind of sacrifice of a bull or ram). These inscriptions advertise the various pagan priesthoods held by the initiate, as well as recording the year and day on which the sacrifice was performed. The earliest altar dates to the year 305, and another to 350, whereas the remaining six belong to the period 374–390 A.D.[6] Moreover, their concentration at this single site indicates that the Phrygeanum, Magna Mater's most important temple in Rome during late antiquity, was located here in close proximity to the church that Constantine had erected over the putative site of St. Peter's grave. This curious topographical juxtaposition, combined with the blatantly pagan character of the taurobolic inscriptions, has been regarded by modern scholars of late antiquity as showing how a number of eminent Roman nobles during the waning decades of paganism publicly broadcast their religious allegiance.[7]

Due to the incompleteness of the ancient evidence, the history and nature of the taurobolium in the cult of Magna Mater are highly problematic. Besides four references to the ritual in surviving literary texts (three of which come from hostile Christian authors), we currently possess 133 inscriptions in which the term 'taurobolium' or 'criobolium' is found.[8] Of these, 128 are associated with the cult of Magna Mater and can be dated no earlier than 160 A.D. The remaining five date to the period ca.135 B.C. to A.D. 134 and are not explicitly connected with the worship of the Great Mother.[9] The four earliest inscriptions come from Asia Minor and allude to taurobolia or criobolia in such a way as to indicate that they were public spectacles, which might be accurately described as ritualistic rodeos in which the animals were chased, somehow overpowered, and then sacrificed. Unfortunately, with very few exceptions, the taurobolic inscriptions relating to Magna Mater offer no precise evidence as to the nature of the ceremony, but characteristic of many epigraphic texts, they do little more than record the performance of the ritual. In fact, in five inscriptions of the taurobolic corpus the mere participial form 'tauroboliatus' is employed in a person's public cursus or epitaph in order to publicize the fact that he or she experienced the ceremony.[10]

According to our most detailed literary source, the Christian poet Prudentius (*Peristephanon* X.1006–1050), who flourished during the second half of the fourth century A.D. and was thus contemporary with the eight inscribed altars from the Phrygeanum in the Vatican, the taurobolium was a kind of bloody baptism. After the initiate stepped down into a pit, a perforated wooden cover was placed over it, onto which a bull was led. The bull's chest was stabbed with a hunting spear (venabulum), and the blood trickled down through the perforations onto the initiate. Furthermore, since the taurobolic inscription of the Roman senator Sextilius Agesilaus Aedesius, dating to August 13 of 376 A.D. (*CIL* VI. 510 = *ILS* 4152), contains the phrase "reborn forever" (in aeternum renatus), many modern scholars have supposed that this rite was thought to confer a blessed afterlife upon the initiate. Accordingly, the initiate's descent into and emergence from the pit would have symbolized death and rebirth.

Generally speaking, modern scholars who have discussed the nature and history of the taurobolium fall into two different groups, who for the sake of convenience can be termed primitivists and minimalists. According to the primitivists the myth of Attis, the youthful herdsman loved by Cybele, who according to one version of his death castrated himself and died shedding his blood beneath a pine tree, is to be interpreted as a myth of a dying and reviving god of vegetation similar to those of Adonis and Osiris. This myth, moreover, was allegorized into the notion of a worshipper's happy life after death long before the cult arrived in Rome in 204 B.C.; and the taurobolium, which somehow formed part of this cult's rites of initiation, remained relatively unchanged throughout its history, being the bloody baptism described by Prudentius. Minimalists such as Rutter and Duthoy, on the other hand, stress the lateness and singularity of the phrase "in aeternum renatus", and they point out that the four earliest inscriptions from Asia Minor which record the term 'taurobolium' or 'criobolium' do not associate the rite with Cybele and are clearly alluding to a ceremony quite different from what Prudentius describes. Consequently, the minimalists argue that over the course of time the taurobolium evolved through different phases, and that Prudentius' bloody baptism was its last phase, which might have been influenced by the Christian concept of baptism. Thus, according to a strictly minimalist interpretation as posited by Lambrechts, the cult of Magna Mater did not actually become a mystery religion until late antiquity.

At first glance the historian might be naturally inclined to side with the minimalists in this matter on the grounds that all institutions and practices are likely to change over time, and one should be careful not to over-interpret the ancient evidence and should avoid anachronism by not attributing later attitudes and practices to earlier times. Nevertheless, it must be emphasized that we are dealing with a very complex phenomenon about which we have very limited information. As a result, no argument from silence can be considered valid. Crucial to a minimalist interpretation is the assumption that our surviving evidence is relatively complete and representative of the entire history of this religious tradition. This assumption, however, is certainly not true; but its unwarranted use by modern scholars in this matter leads to a second methodological fallacy: namely, that the absence of evidence constitutes evidence of absence. We need only consider the following counterfactual proposition in order to understand the hazards posed by arguments from silence in these matters. If both Judaism and Christianity and their literature had not survived beyond late antiquity, how successful would we now be in trying to reconstruct their complex theological systems on the basis of a hundred or so laconic inscriptions and a limited number of brief statements in literary texts written by persons who were either ill informed or hostile to Jews and Christians?

The extreme reticence of the taurobolic inscriptions in regard to a happy afterlife should not cause surprise. As a general rule, epigraphic texts tend to leave many things unstated, partly due to their formulaic language, and partly because they presuppose things about which the ancients themselves were well informed, but which are now unknown to us. In addition, the inscription of Sextilius Agesilaus Aedesius is not the only taurobolic text to allude to afterlife; for two extremely fragmentary inscriptions from Turin in Northern Italy (*CIL* V. 6961–6962) contain the word "aeterni." Unfortunately, these exiguous texts cannot be dated, but they clearly indicate that the Roman senator Aedesius was not the only person who associated the taurobolium with life after death.

The great antiquity of a Cybelic eschatology is further suggested by the surviving ancient evidence concerning the cult of Sabazius. According to Strabo (X.3.15+18) he was a Phrygian deity closely associated with the Phrygian Mother Goddess. Already in the fourth century B.C. his rites were clearly mystic in character, as Demosthenes (*De Corona* 259–260) informs us in a passage in which he heaps scorn upon his political adversary Aeschines by describing how in his youth he assisted his mother in the celebration of these rites. In addition, Demosthenes' quotation of the mystic *symbolon* "I have escaped evil, I have found a better thing" would seem to indicate that the cult of Sabazius resembled contemporary Orphism and the Eleusinian Mysteries in promising its initiates a happy life after death. If so, then the possibility that the Phrygian cult of Cybele and Attis included eschatological beliefs already during the classical period must be taken seriously.

Evidence concerning Greek rites of purification from the shedding of human blood and careful scrutiny of a passage in Herodotus can also be employed to reinforce, if not actually prove, a primitivist interpretation of the taurobolium. Aeschylus in the *Eumenides* (283 and 445–451) indicates that the standard Greek mode of purifying a murderer from the pollution of having shed human blood involved pouring the blood of a sacrificed piglet upon the person and then washing it away with water. A somewhat more detailed description is given by Apollonius Rhodius in his *Argonautica* (IV.703–707). When Circe purifies Jason and Medea, she holds a suckling pig above their heads and cuts its throat, so that the blood falls upon their hands. This agrees with an Apulian red-figure vase painting, depicting Apollo's purification of Orestes, in which the god holds a piglet over Orestes' head.[11]

This ceremony of purification can be connected to the Great Mother Goddess of Phrygia through an anecdote told by Herodotus and attributed to the Lydian royal court of the sixth century B.C.[12] After Solon the Athenian had left the court of King Croesus, who believed himself to be the most fortunate of mortals despite Solon's admonitions, disaster came upon the monarch through the death of his son Atys. After dreaming that his son would die by the blow of an iron weapon, Croesus attempted to avoid this premonition by forbidding his son to engage in warfare, by forcing him to settle down with a wife, and even by having all weapons removed from his son's living quarters. At the same time a Phrygian named Adrastus, a grandson of King Midas, fled to Croesus' court after having killed his own brother. After undergoing purification from the murder according to Lydian custom, which Herodotus says closely resembled that of the Greeks, the man was received among the king's closest associates. Shortly thereafter, the Mysians sought Croesus' aid in hunting down and killing a huge boar that had been ravaging their land. Despite his misgivings, Croesus was persuaded by Adrastus to let him and Atys participate in the boar hunt. When the hunting party finally cornered the beast, Adrastus cast his spear, missed the animal, but struck and killed Atys, thus fulfilling Croesus' dream. Following Atys' burial, Adrastus slew himself upon the grave in despair of having killed the son of his royal host.

As the name of the Lydian prince suggests, this quaint tale is clearly a historicized account of the myth of the Phrygian herdsman Attis, the one beloved of the Great Mother herself. The Lydians and Phrygians were neighbors of one another and shared many cultural traditions, including the worship of the Great Mother of the gods. When the Athenians and Ionians marched inland to attack and burn Sardis, the Lydian capital, in 499 B.C. at the beginning of the Ionian Revolt, they destroyed Cybele's temple, and its destruction was later offered in justification of the Persians' destruction of religious sanctuaries during Xerxes' invasion of Greece (Hdt. V.102). Thus, although Phrygia was traditionally regarded as the homeland of the Great Mother, it is not surprising to find an aspect of her mythology historicized among the Lydians. As already mentioned, in one version of the ancient lore Attis castrates himself and bleeds to death beneath a pine tree, but according to another version he was slain by a boar. It could be that the different accounts of the Attis story represent separate local traditions (e.g., Phrygian vs. Lydian); but in any case, rather than viewing these tales as contradictory, and instead of trying to choose one account over the other, we may be correct in regarding both as having been equally important in forming the mythical basis for the theology of the cult of Cybele. To the mythopoeic mind, as can be seen so clearly from the seemingly contradictory myths of ancient Egypt, variant myths do not so much conflict as they complement one another.

The myth of Attis' self castration was developed to account for the odd custom whereby Cybele's Anatolian priests displayed their devotion to her by castrating themselves and thus empowered the supreme creatrix of the universe by bestowing their virility upon her. Attis' close association with the pine tree stems from the fact that although he was probably regarded originally as the embodiment of the king favored by the Great Mother (Munn 2006 56–130), he was at some point reinterpreted to represent the spirit of vegetation, the male principle complementing Mother Earth; and though most plants grew and died with the changing seasons, the pine tree, plentiful in the mountains of Phrygia, was forever green.[13] Hence the pine tree was integrated into Attis' mythology and was even allegorized to represent the immortal soul triumphing over death. Consequently, the pine cone, a common iconographic element in the art of the cult in Roman times, was probably regarded as a mystic talisman guaranteeing life after death.[14] In addition, the evergreen nature of the pine tree might have been explained by its absorption of Attis' life-giving blood; and this aspect of the myth further explained why in this Anatolian cult human blood had to be shed in order to bring the dead Attis back to life.

On the other hand, Attis' death while hunting a fierce boar was used to explain why worshippers of Cybele and Attis abstained from eating pork (Paus. VII.17.10). The weapon which caused the death of the Lydian prince Atys corresponds to the hunting spear mentioned in Prudentius' description of the taurobolium. Attis' status as a herdsman of sheep and cattle and his involvement in hunting formidable quadrupeds are to be associated with the rodeo-like public spectacles mentioned in the four earliest taurobolic and criobolic inscriptions from Asia Minor. Previous modern critics, such as Rutter and Duthoy, have been mistaken in regarding the taurobolium of the cult of Magna Mater as an outgrowth of these ceremonies, whereas in fact the converse seems closer to the truth. The public spectacles evolved out of a very ancient aspect of Anatolian myth and ritual. As often occurs in the history of religion, the dramatic enactment of a solemn religious tale becomes intertwined with other traditions and gradually evolves into a popular form of entertainment.[15] The earliest criobolic inscription from Pergamum clearly suggests that this local ceremony was the myth of Attis reinterpreted and incorporated into the community's civic traditions: for the chasing and overpowering of the animals were accomplished by the ephebes, whose youthfulness corresponds precisely to that of Attis. Furthermore, the hunting or overpowering of large animals may have been a very ancient and well established rite of passage in Anatolia, whereby a young man demonstrated his entry into manhood; and the story of Attis' death during a boar hunt could be viewed as a mythic variation of this tradition. Moreover, Attis' castration may suggest that the ritualized rodeos grew out of annual activities surrounding the gelding of livestock. How better to demonstrate one's manhood than by overpowering a ram, boar, or bull and then depriving it of its own virility?[16]

The prevalence of Anatolian bull-chasing ceremonies with chthonic and eschatological associations is further suggested by a passage in Strabo and by the Mithraic tauroctony. According to the Greek geographer (XIV.1.44) in a village called Acharaca, situated on the road between Tralles and Nysa, there was a precinct of Pluto and his infernal consort, over which stood a sacred cave. In addition to being the focus of therapeutic incubation rites, the cave figured prominently in an annual festival in which the local ephebes released a bull from the cave, pursued, and somehow killed the animal. It is reasonable to suppose that within this same Anatolian religious tradition arose the Mithraic tauroctony (= 'bull slaying') and its attendant eschatology. Mythological scenes discovered in Roman Imperial Mithraea depict how the cosmic bull escaped from a cave, was chased and apprehended by Mithras, who dragged the animal back into the Mitrhaic cave, where the god knelt upon the animal's back, drew back its head with his left hand while stabbing the animal in the chest with a knife held in his right hand in order to shed its life-giving blood.

To return to the story of the Lydian Prince Atys in Herodotus, the figure of Adrastus represents the inevitability of fate. His name in Greek literally means 'he who does not run away.'[17] Although he flees his native Phrygia after killing his own brother, he cannot escape his fate: for even after he has been purified of this murder, he still ends up committing another homicide by accidentally killing Atys. His name and role in the story are likely to be Greek adaptations of the original Lydian or Phrygian tale, perhaps of Cyzicene origin.[18] In fact, his name might represent a Greek reinterpretation of Agdistis, the name of a hermaphroditic creature found in one version of the Attis tale, an alter ego of Cybele who is responsible for the herdsman's death.[19] Since Adrastus contributed to Atys' death, we would logically expect him to be purified of the blood guilt, but instead, in Herodotus' narrative this purification occurs only at the beginning of the anecdote, not at its end, and is associated with an otherwise extraneous murder. Thus, if the purification in Herodotus has been transposed from the story's end to its beginning in order to illustrate graphically the inevitability of fate, Adrastus in the original tale might have been considered as a kind of 'Everyman', representing each initiate, who according to the theology of the cult must undergo purification from having contributed to Attis' death, a kind of original sin. Consequently, just as the Phrygian worshippers of Cybele and Attis did not eat pork because of the boar's involvement in Attis' death, so the initiate may have originally undergone purification by a shower of a piglet's blood.[20] The myth of Attis would have therefore provided a suitable etiology for the ritual of cleansing someone of having shed human blood. But when in the course of time the boar hunt of the myth was supplanted by ritualized bull and ram chases, the blood of these animals replaced that of the pig; and the rite of initiation, although it merely involved the sacrifice of the animal, retained the name of the ritualized hunt.

Even though certain elements in this reconstruction admittedly rely upon conjecture and speculation due to the nature of the surviving evidence, it is all consistent with ancient modes of mythopoeic thought, and portions of it are sufficiently solid to suggest that a purifying shower of animal blood figured in the cult of Cybele and Attis long before the religion arrived in Rome.

MAGNA MATER'S PLACE IN THE ROMAN STATE RELIGION

A new chapter in the history of the Phrygian cult of Cybele and Attis commenced in 205 B.C. when in response to frequent showers of stones the *decemviri sacris faciundis* were instructed to inspect the Sibylline Books entrusted to their care and announced that according to the sacred verses the foreign enemy on Italian soil (i.e., Hannibal) could be driven out if the Romans imported the cult of the Great Mother from Asia Minor. By enlisting the assistance of King Attalus of Pergamum, the Roman ambassadors received Cybele's sacred stone from her cult center at Pessinus and returned to Italy, where the divinity was solemnly received by the Romans on April 4, 204 B.C., a day henceforth officially commemorated by being the first day of the Megalensian Games (April 4–10) celebrated in honor of the Phrygian goddess, who was henceforth termed Magna Mater by the Romans.[21] Since there was no shrine yet prepared for the new cult, the sacred stone of the goddess was placed in the temple of Victory on the Palatine.[22]

This choice in temporary lodgings was deliberate and laden with symbolic significance. First of all, Magna Mater's cohabitation with Victory was intended to facilitate the fulfillment of the Sibylline prophecy concerning the expulsion of the enemy from Italy. Secondly, her residence on the Palatine, which was also the location of her temple dedicated on April 10, 191 B.C.,[23] was in keeping with the Roman belief that her cult represented their ancestral Trojan religious heritage, because in the Greek mythographic tradition the Great Mother of the gods was closely associated with Mount

Ida in Crete and Mount Ida in the Troad (hence, her epithet 'Idaean').[24] In addition, the name of the Pergamene kingdom ruled by Attalus, who had assisted the Roman ambassadors in their mission, had obvious Trojan associations. Since the Palatine was supposed to have been the original site of Romulus' settlement of Rome and therefore allegedly contained the oldest cults of the state religion, it was deemed only appropriate that the Great Mother of Asia Minor, worshipped by Rome's Trojan ancestors, should have her abode in the most ancient area of the city. Consequently, when Roman senators nearly six centuries later wished to display their opposition to Christianity and their adherence to Roman tradition and paganism, their initiation into the cult of Magna Mater was significant both because of its rivalry with Christianity as a mystery religion, but also due to the fact that it represented Rome's heritage stretching back to Troy. Noteworthy in this regard is the fact that one of the Roman senators initiated into the cult of Magna Mater during late antiquity was Cornelius Scipio Orfitus, whose name suggests his descent from P. Cornelius Scipio Nasica, the man chosen by the senate to receive Cybele's sacred stone when it arrived at the mouth of the Tiber in 204 B.C.[25] Furthermore, in opposition to earlier scholars who have stressed Julian's involvement in Mithraism, R. Smith (1995 178) has argued that the cult of Magna Mater figured prominently in the emperor's avowed paganism and opposition to Christianity.

Rome's religious history had always been characterized by the importation of new cults and practices in order to serve the needs and interests of the Roman people. For example, following the conquest of Veii in 396 B.C. Rome had appropriated unto itself the divine favor of its erstwhile Etruscan rival by installing the Veientine cult of Juno Regina on the Aventine; and the periodic occurrence of plagues during the early Republic had resulted in the introduction of a cult to Apollo Medicus in 431, the celebration of the first lectisternium in 399, the importation of stage performances from Etruria in 364, and the introduction of the cult of Aesculapius from Epidaurus in 292 B.C.[26] Nevertheless, no other cult had been so alien to Roman religious traditions as was that of Cybele, whose worship consisted in part of wild dancing and howling, accompanied by the frantic music of cymbals, drums, and flutes, as well as the participation of castrated priests robed and adorned like women, who amid religious frenzy shed their own blood by self flagellation. It appears that Roman officialdom had responded to the Sibylline Books without being fully aware of the exotic nature of the Phrygian cult. The Romans, however, did not repudiate the Great Mother; they simply Romanized her worship by establishing yearly games and aristocratic banquets in her honor and by allowing her imported Anatolian ministrants to observe their traditional religious practices among themselves under the watchful eyes of Roman officials. Thus, just as the worship of Demeter had been introduced from the Greeks of southern Italy and Sicily into Rome in the form of the cult of Ceres on the Aventine during the early fifth century B.C., and her cult was always served by a priestess from a Greek city while still being part of the Roman state religion, so the cult of Magna Mater included foreign ministrants but also formed part of the official religion.[27] Inscriptions of Imperial date clearly show that the cult always remained under the supervision of the *quindecemviri sacris faciundis*, the board of priests responsible for the cult's arrival in Rome in 204 B.C. through their consultation of the Sibylline Books. Thus, in comparison to some other mystery religions, the Phrygian cult throughout the remainder of its history enjoyed a somewhat unusual and privileged status. It was permitted to observe unofficially its native traditions while officially receiving Roman religious accretions and the general imprimatur of the Roman state; and this peculiar symbiosis, combined with the Roman belief in the cult's ancestral Trojan character, ultimately had important ramifications that have hitherto not been fully appreciated by modern scholars.

As the result of Rome's position as an Imperial power, Roman society became increasingly diverse and was permeated with Hellenistic religious ideas during the last two centuries B.C. Nevertheless, since the official Roman state religion was generally more conservative than the religious beliefs and behavior of many inhabitants of Rome, a divide arose between the official religious policy of the state and the unofficial religious practices of numerous individuals. Magna Mater's unusual place in the religion of the Roman state becomes apparent when one recalls how some other cults from the Hellenistic East were rudely handled by Roman officialdom during this period. Eighteen years after the cult of Phrygian Cybele came to Rome, and five years after her temple on the Palatine was dedicated, the Roman state in 186 B.C. proceeded very vigorously and forcefully to impose stringent regulations upon Bacchic worship throughout Italy and to punish severely those who were in violation of the expressed wishes of the Roman senate.[28] In 139 B.C., astrologers and Jews were banished from Rome (Val. Max. I.3.2 and Livy *Per. Oxy.* 54). Despite its popularity among slaves, freedmen, foreigners, and even Roman citizens, Roman officials took repressive measures against the worship of Isis during the last years of the Republic; and Octavian's victory over Antony and Cleopatra at Actium in 31 B.C. insured that Roman official antipathy toward the gods of Egypt continued into the early principate.[29]

In describing the institutions established by Romulus, Dionysius of Halicarnassus (II.19), writing during the Augustan principate, uses the cult of Magna Mater to illustrate the contrast between the sedate nature of Roman religious practices and the much more ecstatic ones of the Hellenistic religions. He observes that the Romans have no such traditions as Uranus' castration by Cronus, nor the latter's attempt to destroy his children, nor Zeus' dethroning of his father and the war among the gods, nor the servitude of deities among men. The Romans do not have festivals characterized by mourning, the wearing of black garments, the beating of breasts, and the lamentation for the disappearance of divinities such as Persephone or Dionysus. Nor do Roman rites contain ecstatic transports, Corybantic frenzies, sacred begging, Bacchanals, secret mysteries, or all-night vigils. Indeed, even when the Romans have adopted foreign cults in accordance with oracles, they have observed them with ceremonies consistent with their own traditions, while excluding the exotic claptrap of foreign rites. Thus, for example, although the Romans have received the cult of the Idaean Mother and have allowed her to be served by a Phrygian priest and priestess, who bear her image through the city with their customary begging, lamentation, and music of flutes and cymbals, the praetor conducts sacrifices and games in her honor, and free-born Roman citizens are forbidden by decree of the senate to participate in the Phrygian ceremonies.[30]

The tenor of Dionysius' remarks strongly implies that the native Phrygian cult of Cybele and Attis during his own day was both orgiastic and mystic, and that the Romans during the last two centuries B.C. did not officially incorporate these features into the state religion. The references to Cronus' castration of Uranus, the frenzied dancing of the Corybantes, the disappearance of deities such as Persephone and Dionysus, sacred begging, and the observance of secret mysteries and vigils all allude directly or indirectly to the Phrygian cult of Cybele and Attis, whose contrastingly sedate worship by the Romans is the subject of the remainder of this illustrative digression. Thus, this chapter of Dionysius' *Roman Antiquities* not only tells us how the Romans in the Augustan age officially worshipped Magna Mater, but it also constitutes important, albeit indirect, evidence for the mystic character of the Phrygian cult during the early years of the principate.

The Roman state's aloofness toward Eastern cults began to relax with the death of the Emperor Tiberius, who had strictly adhered to the policies of Augustus throughout his reign. Tiberius' immediate successor, Gaius Caligula, may have initiated the "Orientalizing" of the Roman state religion by establishing a temple to Isis in the Campus Martius.[31] The full acceptance of the Egyptian cult into the Roman religious experience was evident when the Emperor Vespasian, a man from the traditionally conservative Sabine Territory of central Italy, numbered himself among its devotees.[32] As suggested by Juvenal's famous gibe concerning the Syrian Orontes pouring its culture into the Tiber (*Sat.* III.62–65), the absorption of Hellenistic religions into Roman society continued apace, and its effects are perhaps seen most strikingly in the "Great Bacchic Inscription of the Metropolitan Museum," a marble statue base inscribed on three sides with the names and Bacchic titles of nearly 400 members of the religious association, including many prominent individuals, about one third of whom were women. The statue was erected in honor of Pompeia Agrippinilla, the priestess of the organization. The monument dates to the middle of the second century A.D. and was discovered in Latium.[33] Three centuries earlier the Roman senate had imposed the death penalty upon persons in Italy who contravened the senate's regulations of Bacchic worship, but such senatorial strictures on Dionysiac religion were totally unthinkable during the age of the Antonine emperors.

The absorption of Hellenistic religions into Roman society during the first and second centuries A.D. paved the way for the Roman state's official acceptance of the Phrygian mystic rites surrounding the death and resurrection of

Attis. According to the Calendar of Furius Philocalus, dating to the year 354 A.D. (Degrassi 1963 242–243), the official Roman calendar of late antiquity contained the following series of festivals for the cult of Magna Mater:

March 15: Canna Intrat = The Reed Enters
March 22: Arbor Intrat = The Tree Enters
March 24: Sanguem = Blood
March 25: Hilaria = Rejoicing
March 26: Requietio = Rest
March 27: Lavatio = Bathing

The first day presumably symbolizes Attis' birth and exposure among the reeds alongside the River Sangarius in Phrygia. Seven days later (an astrological week possibly representing Attis' mortal existence), the tree-bearers (dendrophori) of Magna Mater's cult cut down a pine tree and bore it into the temple amid lamentation and mourning for the death of the Phrygian herdsman, whose image was suspended from the branches of the tree. Thus began a period of three days of grieving observed by the worshippers. The third day must have culminated in frantic exhibits of sorrow, including the cutting of one's flesh and the shedding of blood.[34] This in turn apparently brought about Attis' rebirth on March 25, the vernal equinox (Macrob. *Sat.* I.21.10), the first day of spring, thus ushering in the Hilaria to celebrate the god's miraculous revival. It is noteworthy that Christians from the early third century onwards associated this same day with Jesus' resurrection (Tertullian *Adversus Iudaeos* 8 and Lactantius *De Mortibus Persecutorum* 2.1). The ceremonies were concluded by one day of rest and another on which the sacred statue and other utensils of the Great Mother were carried in procession from the Palatine temple to the Almo, a stream that flowed into the Tiber, where the sacred objects were solemnly washed and were brought back to the temple.

According to IV.59 of the *De Mensibus* of Johannes Lydus, a Byzantine writer of the sixth century A.D., the Emperor Claudius introduced the Dendrophoria (= Arbor Intrat of March 22) into the Roman calendar. Lydus' statement receives support from Suetonius, who in his list of omens portending the Emperor Otho's death in 69 A.D. includes the fact that the emperor left Rome to encounter Vitellius in northern Italy on the day on which the worshippers of Magna Mater begin their mourning (*Otho* 8.3). Modern scholars have generally accepted Lydus' attribution of the Dendrophoria to the Emperor Claudius as true, but disagreement has arisen with respect to the introduction of the other March festivals. Carcopino (1941 51–59) argued that Claudius was responsible for the entire series of festivals, whereas other scholars more recently have favored the idea that they were gradually admitted into the Roman state calendar, beginning with the reign of Claudius and probably ending in the time of Antoninus Pius.[35] As in the case of a minimalist approach to the history of the taurobolium, scholars who propound an evolutionary thesis for the incorporation of these March festivals into the official Roman calendar rely upon shaky arguments from silence. Although epigraphic and numismatic data indicate that Antoninus Pius promoted the cult of Magna Mater, both literary and artistic evidence suggests that the March ceremonies were already fully incorporated into the official Roman calendar during the first century B.C, thus rendering Carcopino's Claudian thesis entirely plausible.

Given Lydus' rather desultory and haphazard manner of selecting antiquarian information for inclusion in his treatises, his ascription of the Dendrophoria alone to Claudius and his reticence concerning the other March festivals cannot be regarded as a valid argument from silence.[36] The *Epigrams* of Martial (III.47.1–2 and XI.84.1–4) and a passage in the *Argonautica* of Valerius Flaccus (VIII.239–241) demonstrate that the Lavatio of March 27 was well established in Rome during the Flavian period. Moreover, both authors' allusion to the shedding of blood likewise indicates that the *Dies Sanguinis* of March 24 was also well known during their day. Some of the scenes depicted in the ceiling of the Basilica of the Hypogeum near the Porta Maggiore in Rome, discovered on April 21, 1917, indicate that as early as the first half of the first century A.D. Romans associated Attis with death and rebirth. The four corners portray Attis in mourning, but the central panel shows a winged Attis holding a torch and conducting Ganymede to heaven.[37] This version of Ganymede's apotheosis involving Attis seems to have been current before Strabo's day, because the Greek geographer (XIII.1.11) records that according to some writers Ganymede's abduction occurred at a site appropriately called Harpagia in the Plain of Adrasteia, an area closely associated with the myths of Cybele and Attis. The scenes in the Hypogeum of the Porta Maggiore are consistent with the Hilaria that celebrated Attis' resurrection.

The introduction of the March ceremonies of Magna Mater and Attis into the official Roman calendar harmonizes with the thought and actions of the Emperor Claudius. As the result of his serious study of history (Suet. *Claud.* 41–42), many decisions and policies of the emperor can be viewed as innovations that were considered to be in accord with Rome's historical traditions. The best illustration of this phenomenon is Claudius' well-known enrollment of notables from Gallia Comata into the Roman senate, which the emperor justified by appealing to similar innovations throughout Roman history (*ILS* 212). Claudius' respect for Rome's Trojan ancestry and religious heritage are evident from two items recorded by Suetonius (*Claud.* 25.3–5). The emperor granted the people of Ilium immunity from taxation due to their kinship with the Roman people, and in support of his decision he adduced an old letter of the Roman people and senate to King Seleucus. The emperor also rebuilt the temple of Venus on Mount Eryx in Sicily, which according to Roman belief had originally been established by Aeneas himself (Dion. Hal. I.53.1, cf. Diod. IV.83). It is therefore reasonable to suppose that Claudius took a keen personal interest in the cult of Magna Mater due to the fact that a member of his own patrician family, Quinta Claudia, was thought to have played an important role in the cult's introduction to Rome in 204 B.C. Both Livy and Ovid indicate that the story of her allegedly miraculous dislodging of Magna Mater's ship from a sandbar in the Tiber and the concomitant vindication of her chastity was well known during the Augustan age, probably because it formed part of a religious drama performed every year during the Megalensian Games of April that retold the story of how the Great Mother of the gods came to Rome from Asia Minor.

Two other items recorded by Suetonius (*Claud*. 20.3 and 25.5) may further indicate Claudius' interest in the rites and ceremonies of the Phrygian cult. In cataloguing the emperor's sponsorship of public spectacles, the imperial biographer informs us that Claudius introduced a new kind of entertainment in which cavalrymen pursued wild bulls in the circus, jumped upon them when they were worn out, and dragged them to the ground by their horns. This spectacle is clearly the public taurobolium known from four inscriptions from Asia Minor, one of which comes from Troy itself. Secondly, in describing various actions taken by the emperor in the sphere of religion, Suetonius remarks cryptically, "he even tried to transfer the Eleusinian Mysteries from Attica to Rome." Exactly what lies behind this curious statement cannot be understood in the present state of our knowledge, but it may at least indicate Claudius' serious interest in mystery religions; and if he in fact ever seriously considered introducing the prestigious Eleusinian Mysteries into Rome only to abandon the idea in respect for their historic association with Eleusis, he could have substituted in their place the mystic rites of Cybele and Attis, whose introduction could have been justified by claiming the ceremonies to be of Trojan origin.

In addition to the tradition of Quinta Claudia, the emperor's interest in the Phrygian cult is likely to have been further stimulated by another bit of family religious history. In 296 B.C., Ap. Claudius Caecus vowed a temple to Bellona during a battle with the Etruscans and Samnites (Livy X.19.17 and *ILS* 54). The temple's *dies natalis* was June 3 (Ovid *Fasti* VI.201–204); and it stood in the Campus Martius near the Circus Flaminius and throughout the Republic was often the place where the senate convened to receive commanders in the field or foreign ambassadors. The shrine was decorated with *imagines clipeatae*, shields that bore the portraits of famous members of the Claudian family accompanied by inscriptions that enumerated their public offices and achievements (Pliny *NH* XXXV.12). Thus, the edifice was not only a well-known public shrine, closely associated with important affairs of state, but it was also a monument that celebrated the patrician Claudii, a fact of which the emperor must have been well aware. By the late Republic the Roman goddess Bellona was equated with the Cappadocian deity Ma, who in turn was identified with Phrygian Cybele.[38] Thus, Ma Bellona and Magna Mater were thought to be one and the same divinity. If this syncretism existed as early as the Hannibalic War, it would explain why the Romans sought the aid of the Mother of the gods in 205 B.C. to drive an enemy army from Italian soil. Claudius' interest in the cult of Bellona is suggested by an inscription recording the dedication of a lamp to the goddess by one of his slaves. The dedicatory plaque was set up in Samnium among the Ligures Baebiani and dates to the year 11 A.D. when the future emperor was only about 20 years of age (*ILS* 3806).

Two inscriptions may offer additional support to the notion that Claudius was instrumental in incorporating the March festivals of Cybele and Attis into the Roman calendar. The text of *ILS* 4164, found somewhere in Rome and now known only from a modern transcription, records the epitaph of a man named Ti. Claudius Velox, who bears the title "foremost public eulogist of the Idaean Mother of the gods and of Attis = hymnologo primo M(atris) d(eum) I(daeae) e[t] Atti[n]is publico." The adjective 'publico' clearly indicates that the office recorded in this inscription was part of the Roman state religion. The form of the man's name, consisting of the *praenomen* and *nomen* of the Emperor Claudius, suggests that the deceased himself or an ancestor had been either a peregrine enfranchised by the emperor or one of his freedmen; and the title 'hymnologus' of Magna Mater and Attis is much more in accord with the Phrygian mystic rites of March than the Roman-style Megalensian Games of April. Moreover, the adjective 'primo' could mean that Ti. Claudius Velox was the first man who held this position, but since other inscriptions demonstrate that some local cults of Magna Mater possessed at least two grades of priests and musicians,[39] it is probably better to interpret 'primo' in similar hierarchical terms. Thus, if this terse epitaph could be dated to the Julio-Claudian period, it would indicate that Claudius established at least two grades of *hymnologi* whose function was to narrate the mythology and to praise the majesty of Magna Mater and Attis in public during their ceremonies in March.

The other epigraphic text that has a bearing upon the Emperor Claudius' possible involvement in the mystic rites of Cybele and Attis is *CIL* X. 1596, one of the five taurobolic inscriptions that does not explicitly mention Magna Mater. It comes from Puteoli and dates to the year 134 A.D. It records Herennia Fortunata's repetition of an *ecitium taurobolium* of Venus Calestis and a *pantelium*, performed with the assistance of a priest named Ti. Claudius Felix. Vermaseren (1977 102) is certainly right in conjecturing that the Venus Caelestis of this text is none other than Cybele herself, who has been equated with the tutelary goddess of Carthage. Thus, if we leave aside the four taurobolic inscriptions from Asia Minor that refer to rodeo-like public spectacles, this text constitutes the earliest datable performance of a private taurobolium. The priest mentioned in the text must have been an official in the local cult of Magna Mater. As in the case of the inscription discussed in the preceding paragraph, if the *nomen* and *praenomen* are taken to be of libertine origin, Felix's association with the Great Mother of the gods could have been acquired through his family's connection to the imperial house. In any case, ancient *testimonia* from different quarters converge in providing strong circumstantial evidence that the Emperor Claudius was interested in the cult of Magna Mater and was probably responsible for introducing into the official

Roman calendar the entire set of March ceremonies recorded by Furius Philocalus in his calendar of 354 A.D.

The inscription from Puteoli has additional significance for the history of the taurobolium in the cult of Magna Mater. Since it precedes the reign of Antoninus Pius, it clearly demonstrates that the latter emperor was not responsible for incorporating this rite into the Phrygian cult as has been sometimes supposed. Rather, as already argued in the first part of this chapter, the taurobolium had long been an integral part of the Phrygian rites. Antoninus Pius' involvement in the cult probably lay in the construction of the Phrygeanum in the Vatican (which henceforth replaced the Palatine temple as Magna Mater's chief shrine in Rome) and the adaptation of the taurobolium to form a new element in the cult of the Roman emperors. Thus, from his reign onwards there existed two kinds of taurobolia: the mystic and the imperial, the former being the traditional private rite of initiation, and the latter being the new type of taurobolium performed for the sake of the emperor's welfare (pro salute Imperatoris).[40] Accordingly, the epithet 'Salutaris' bestowed upon Magna Mater on the coins of Antoninus Pius clearly commemorates this innovation and advertises the fact that the Great Mother of the gods was now regarded as having extended her protective powers over the emperor. This innovation represents nothing more than the Romans' continuing interpretation of the historical and religious significance of their Trojan ancestry. Hadrian had recently erected a magnificent temple to Venus and Roma on the Velia, which memorialized the Romans' descent from the Trojan Aeneas and his divine parent. Similarly, the creation of the imperial taurobolium and the construction of the Phrygeanum are likely to have been inspired by the celebration of Rome's 900th anniversary during the reign of Antoninus Pius.

DIFFUSION OF MAGNA MATER'S CULT IN THE LATIN WEST

Given the unique position that Magna Mater's cult enjoyed within the Roman state, it should come as no surprise that the cult's diffusion was inextricably bound up with the Romanization of the provinces and was therefore confined to the non-Greek areas of the empire. The Phrygian cult was only one of many other institutions that communities in Gaul, Spain, and North Africa adopted during the first two centuries of our era to demonstrate their membership in the Imperial system. The great success of Romanization in the Latin West is revealed by a passage of Aulus Gellius' *Attic Nights* (XVI.13), a chapter that discusses the difference between a *municipium* and a *colonia*. As the name suggests, the latter was a community established by the Roman state, whose laws and institutions prevailed in the colony, whereas a *municipium* was an independent foreign state that had been granted Roman citizenship but continued to use its own indigenous laws and institutions. Generally speaking, during the late Republic and early principate *municipia* were content to enjoy the rights and privileges of Roman citizenship while at the same time they jealously preserved and maintained their own local traditions. By the middle of the second century A.D., however, as Gellius indicates, the prestige of Rome had progressed so far that many municipia petitioned the senate and emperor to grant them the legal status of a colony so as to be regarded as replicas of the Imperial capital.

As is evident from the abundant epigraphic and artistic data collected by Vermaseren in the volumes of *CCCA*, Magna Mater's cult was ubiquitous throughout Italy and the Western provinces, and the spread of the cult was already well under way during the first century A.D.[41] One inscription unearthed at Herculaneum (*ILS* 250) shows that the local temple to Magna Mater was damaged by an earthquake and was restored by the Emperor Vespasian. Another inscription (*IRT* 300), dating to the reign of the same emperor, testifies to the existence of a temple to Cybele and Attis in Lepcis Magna in North Africa. An epitaph from a small town near Massilia in Gallia Narbonensis (ILS 4100) records that the deceased had been an apparitor of Magna Mater Palatina. The epithet 'Palatina' could mean that the man had served the Phrygian cult in Rome on the Palatine, but it seems more likely that the adjective refers to a local cult and signifies its imitation of the one in Rome. Similarly, *ILS* 3805 from Kastel near Mainz on the Rhine records that in 236 A.D. a local association of hastiferi of the Great Mother was responsible for restoring a Mons Vaticanus, a structure that was clearly in imitation of the Phrygeanum in Rome. Two other documents record divine epithets of the goddess that testify to the close connection between local cults and the one in the Imperial capital. *CIL* X. 4829 from Rufrae in central Italy attributes to the Mother of the gods the epithet 'Optima Maxima' clearly in imitation of Jupiter Optimus Maximus, the chief tutelary deity of Rome. Another text from a small town in Narbonensis adds to the official Roman title Mater Deum Magna Idaea the adjectives 'Phrygia Palatina'.[42]

The Phrygian cult must have been quite attractive in that it offered a local community a direct connection to Rome and its venerable Trojan ancestry as well as providing the local citizenry with the services of a mystery religion. The inscriptions indicate that the local cults comprised numerous functionaries (archigalli, sacerdotes, cannophori, dendrophori, hastiferi, tibicines, cymbalistriae, tympanistriae, apparitores), thereby allowing devotees to serve the goddess in a wide range of official capacities.[43] Women are well represented in the epigraphic texts. Thirteen inscriptions in the taurobolic corpus contain the names of female *sacerdotes*. In fact, a document from Beneventum (*CIL* IX. 1541) records a taurobolium experienced by a priestess of the second order that was supervised by a priestess of the first order. In listing the participants and witnesses of taurobolic ceremonies, two inscriptions from Mactar in North Africa (*CIL* VIII. 23,400–401) contain the phrase "together with all the dendrophori and the devotees of both sexes (una cum universis dendrophoris et sacratis utriusque sexus)."

The epigraphic evidence indicates that priests of local cults of Magna Mater often bore the title 'sacerdos quindecemviralis'.[44] Three inscriptions from Ostia (CIL XIV. 40, 42, and 4303) record the performance of a taurobolium not only for the welfare of members of the imperial family, but also for other politically and socially significant groups in Roman society, listed in descending order of perceived importance. The first two groups following the imperial family are the senate of Rome and the college of *quindecemviri sacris faciundis*. An inscription from Baiae (*ILS* 4175), dating to the year 289 A.D. and concerning the appointment of a local priest of Magna Mater, provides specific information on the relationship between the *quindecemviri* of Rome and the local cults of the Idaean Mother throughout Italy and the provinces. The first part of the text records the local town council's choice of a priest to fill the vacancy caused by death, and the second part of the text sets forth a letter from the quindecemviral college of Rome, confirming the appointment.

> On June 1 of the consulship of | M. Magrius Bassus and L. Ragonius | Quintianus in Cumae in the temple of the deified Vespa- | sian in a meeting of the town council, | which the praetors M. Mallonius Undanus | and Q. Claudius Acilianus | had summoned, there were present at the writing the following members | chosen by lot: Caelius Pan- | nychus, Curtius Votivus, and Considi- | us Felicianus.[45] When the praetors brought forward the matter | of choosing a priest for the Baian Mother | of the gods in place of the deceased | priest Restitutus, it was unanimously decreed | that Licinius Secundus | be chosen priest. |
>
> The board of fifteen for performing rites sends greetings to the praetors and magistrates of Cumae. | Since we have learned from your letter | that you had appointed Licinius | Secundus priest of the Mother of the gods | in place of the deceased Claudius Restitutus, | we grant him permission | in accordance with your wish to wear the bracelet | and crown within the | confines of your colony. | We hope that you are well. | I Pontius Gavius Maximus, the vice chairman, have endorsed with my signature on Aug. | 17 in the consulship of M. Umbrius Primus | and T. Flavius Coelianus.[46]

Thus, from this document it emerges that the *quindecemviri sacris faciundis* of Rome, the priestly college originally responsible for the introduction of Magna Mater from Asia Minor in 204 B.C., continued their nominal supervision of the cult well into Imperial times by ratifying the appointment of local priests. The inscription also suggests that these officials wore a standard religious habit, consisting in part of a crown and a bracelet. Hence, we may safely conclude that these local cults and priesthoods of Magna Mater generally conformed to a formal religious canon as established and followed by the cult in Rome itself. The epigraphic evidence indicates that these priests presided over both mystic and imperial taurobolia.

THE IMPERIAL MYSTERIES OF MAGNA MATER

The mystic and official aspects of Magna Mater's Phrygian cult interacted and influenced one another in ways that have not been fully appreciated by modern scholars. The one document that is most revealing in this regard is an inscription from Lugdunum dating to the year 160 A.D. that reads as follows:

> Having performed the taurobolium of the Great Idaean Mother of the gods | in accordance with the commandment of the Mother of the gods | for the welfare of the Emperor Caesar T. Aelius | Hadrian Antoninus Augustus Pius, father of his country, | and for the welfare of his children | and of the state of the colony of Lugdunum, | L. Aemilius Carpus, a sevir Augustalis and | a dendrophorus, | [carving of a bull's head] | received the *vires*, transferred them from the Vatican, | and at his own expense consecrated | an altar and a Bucranium in the priesthood of | Q. Sammius Secundus, ordained by the board of fifteen | with the bracelet and crown, | to whom the most honorable town council of Lugdunum | has decreed the priesthood for life, | in the consulship of Appius Annius Atilius Bradua and T. Clodius Vibius | Varus, | a site having been made available by decree of the town council.
>
> [Inscribed on the right side of the stone]: whose *mesonyctium* | was performed on December 9.[47]

This taurobolium, performed for the welfare of Antoninus Pius, his children, and of Lugdunum itself, was undertaken by L. Aemilius Carpus, who describes himself as a sevir Augustalis for the imperial cult at Lugdunum as well as a dendrophorus in the local cult of Magna Mater. At his own expense Carpus erected an altar and a Bucranium for the cult of the Great Mother on a site provided by the local town council and with the supervision of the local state priest of the Great Mother of the gods. Many taurobolic inscriptions, such as this one, were carved on the consecrated altars commemorating the taurobolium. An inscription from Diana Veteranorum in southern Gaul (*CIL* XII. 1567 = *ILS* 4140) contains the phrase "the *vires* were deposited at the spot (= loco vires conditae)," thus indicating that the testicles of the sacrificed animal were solemnly deposited beneath the altar. The epigraphic texts from two taurobolic altars from Lactora in Aquitania (*CIL* XIII. 522 and 525) demonstrate that the *vires* were consecrated together with the altar. Vermaseren (1977 111), citing a *scholion* on Apollonius Rhodius' *Argonautica*, explains this curious custom as deriving from the native Phrygian cult as observed among the Cyzicenes, among whom persons who had castrated themselves in service to Cybele buried their amputated testicles in a sacred cave.

The inscription agrees with the previously quoted document from Baiae in showing that the local priest of Magna Mater received his lifetime office and right to wear the cult's sacerdotal bracelet and crown by vote of the local town council with the endorsement of the *quindecemviri sacris faciundis* in Rome, But two other details make this taurobolic inscription unique and worthy of closer examination. First, this taurobolium was apparently performed at the Phrygeanum in the Vatican in Rome, not in Lugdunum, and Carpus brought the bull's *vires* back to Lugdunum with him. This detail testifies to the close connection between the cult of the Phrygeanum in Rome and those throughout Italy and the provinces. The rite was performed "in accordance with the commandment of the Mother of the gods (ex imperio matris deum)," suggesting that it was prompted by a dream, probably experienced by Carpus in Lugdunum, after which he sojourned to Rome to fulfill the divine injunction.[48] Secondly, the two-line relative clause inscribed on the right side of the stone indicates that Carpus also engaged in a rite termed mesonyctium, and that this ceremony was performed on December 9. Since this clause has been cut on the right side of the stone at a level that is just below the first three lines of the main inscription which record the performance of the taurobolium in accordance with the commandment of the Mother of the gods, the grammatical antecedent of the relative clause must therefore be *taurobolio*. This interpretation suggests that at least in some instances a taurobolium was accompanied by the observance of a mesonyctium.

Although the Latin word 'mesonyctium' occurs nowhere else except in this document, its obvious Greek origin makes it clear that it was some sort of ritual associated with midnight. Evidence from other mystery religions can help to establish more precisely the nature of this ceremony. Lucius in Apuleius' *Metamorphosis* XI.23 experienced his initiation into the mysteries of Isis at midnight: "I approached the region of death. After trodding on Proserpina's threshold, I was borne through all the elements and came back again. At midnight I beheld the sun shining and blazing with light. I came into the presence of the nether and heavenly gods and worshipped them close at hand." The Eleusinian and Orphic mysteries also seem to have been conducted at night time and probably had their crucial moments of mystic revelation illuminated by blazing lamps or torches. This ceremony, termed 'pannychis', and literally meaning 'all nighter', is clearly analogous to the word 'mesonyctium'. Thus, the latter term should refer to a nocturnal ritual that most likely included the enactment of a divine drama involving those being initiated into the mystic rites of Cybele and Attis, whose culminating revelation occurred at midnight. Since the promise of a happy existence after death was the central feature of the mystery religions, the dramatic culmination of the mystic rites of the Phrygian cult is likely to have involved Attis' miraculous resurrection. Judging from the list of the Phrygian ceremonies in March recorded in the calendar of Furius Philocalus, it is reasonable to suppose that the mesonyctium was normally observed by the archigallus at the Phrygeanum in Rome during the evening of March 24, the day of blood, because it then would have ended dramatically at midnight, thereby ushering in the Hilaria of March 25, the day on which Attis' resurrection was celebrated. This dating of the mesonyctium further reinforces its connection with the taurobolium, because the latter ritual, involving the shedding of a bull's blood, agrees with the bloodshed of March 24.

A Greek inscription, discovered at Thessalonica and containing the word 'mesonyction' in Greek, may offer an additional clue to the nature of the Phrygian midnight ceremony. The document records a gift of land to a private mystic association of the god Zeus Dionysus Gongylus (*CRAIBL* 1972 pp.478–479). The *mystai* of the cult are enjoined to use the produce of this land for three annual banquets (whose calendrical dates are indicated) and to observe "the mesonyction of bread." The latter phrase demonstrates that in this particular religious club the midnight ritual involved the ceremonial consumption of bread. The worshippers of Cybele and Attis observed a nine-day period of abstinence from wine and bread, beginning on March 16, the day after the Cannophoria.[49] This nine-day period would have therefore ended at midnight on March 24/5. Moreover, Firmicus Maternus (*De Errore Profanarum Religionum* 18.1) records the following mystic *symbolon*: "I have eaten from the drum; I have drunk from the cymbal; I have mastered the secrets of religion." Since the drum and cymbal were characteristic of the cult of the Great Mother, this utterance must derive from the initiation rites of the Phrygian cult; and the wording clearly suggests that the worshippers ended their abstinence from wine and bread by participating in a sacramental meal in which the wine and bread were consumed from the sacred objects of the cult. Thus, the Hilaria, commencing just after midnight on March 25, would have combined the rejoicing over Attis' resurrection with the good cheer of communal dining, whose appreciation was doubtless enhanced by the nine days of abstention and by the physical ordeal of March 24: for the Day of Blood most likely involved wild dancing and even the cutting of one's flesh to shed blood in order to revive the god. Consequently, the Phrygian mesonyctium, it would appear, was a solemn nocturnal vigil that included fasting and the enactment of a ritualistic drama that culminated at midnight with an impressive mystic revelation under dazzling torch light, probably accompanied by beating drums, clashing cymbals, and trilling flutes, followed by a celebratory mystic meal. Thus, the fortuitous inclusion of the word 'mesonyctium' in *CIL* XIII. 1751 makes an important contribution to our knowledge of the rites surrounding the death and resurrection of Attis.

Nevertheless, it should be recalled that L. Aemilius Carpus performed his taurobolium and observed the mesonyctium on December 9, not March 24. Following the bloodshed of the taurobolium, Carpus must have observed the mesonyctium during the evening of December 9. Thus, the midnight implied by the term 'mesonyctium' must have been that of December 9/10. From early Republican times onwards, December 10 was the day on which the plebeian tribunes entered office. This date was obviously chosen because in early times December was the tenth month in the Roman year, and the Romans must have considered it appropriate, if not actually auspicious, for the ten plebeian tribunes to enter office on the tenth day of the tenth month. Although the plebeian tribunate became a minor administrative office under the principate,

Augustus turned its populist associations with the Roman people to his own political advantage and made its power, the *tribunicia potestas*, a key element in his reorganization of the Roman state and his constitutional position therein. From 23 B.C. onwards he refused to be elected to consecutive consulships as had happened since the year of Actium; and rather than advertising his proconsular *maius imperium* and his extraordinary *provincia* that gave him control of the military forces throughout the whole empire, he prominently displayed his tenure of tribunician power, which in the words of Tacitus (*Ann.* III.56) he transformed into "the title of the highest eminence (= summi fastigii vocabulum)."[50] Henceforth, he and subsequent emperors included their tribunician power in their official titulature and used it to mark off the years of their reigns. From the time of Trajan onwards, however, December 10 was the day on which each new year of an emperor's tribunician power began.[51] Consequently, Carpus' observance of the mesonyctium in 160 A.D. was intended to commemorate in mystic terms the renewal of Antoninus Pius' imperial authority.

This imperial mesonyctium can best be understood in the context of what H. W. Pleket (1965) has called "Imperial Mysteries." After collecting and discussing in detail the exiguous and often fragmentary epigraphic evidence from Asia Minor and Bithynia, dating to the second century of our era, that contained titles such as 'sebastophantes' and 'sebastologos' (which combine the terminology of emperor worship and religious mysteries), Pleket argued that these data pointed to the celebration of religious mysteries in which imperial insignia and icons were the focus of mystic rites of revelation. Pleket's study suggests that two apparently unrelated modes of religious thought, mystery religions and ruler worship, could freely interact to produce imperial mysteries. The taurobolic inscription of L. Aemilius Carpus from Lugdunum suggests that such imperial mysteries were not confined to the Greek East. The significance of the calendrical date December 9/10 for the renewal of the emperor's tribunician power renders it highly probable that the mesonyctium observed by Carpus culminated in a revelation that somehow involved imperial insignia and/or icons. As already observed, much of our information on the cult of Magna Mater comes from rather uninformative inscriptions that usually record little more than the performance of a taurobolium, whether imperial or mystic. The fortuitous preservation of this inscription with the calendrical date December 9 and the term mesonyctium raises the possibility that many other, less informative documents recording taurobolia performed for the emperor's welfare involved imperial mysteries as well. Moreover, since the cult's official propagation seems to have been supervised by the *quindecemviri sacris faciundis* of Rome, who even numbered the emperor himself among their members, it is possible that the same priestly board could have been at least partly responsible for the creation or the diffusion of imperial mysteries that were attached to the cult. Thus, just as Augustus revived the archaic rites of the *Fratres Arvales* and wedded them to the new political regime by having them performed largely *pro salute imperatoris*, we should not be surprised to find that during the reign of Antoninus Pius, when the mystery religions had become widely accepted throughout the empire, the mystic rites of Magna Mater were adapted to the same end.

So much for viewing the inscription from the perspective of Roman officialdom. It may also be worthwhile to try to view it from the perspective of a devout ancient worshipper. Ruler cult seems to have arisen spontaneously in Asia Minor during the fourth century B.C. from a peculiar mindset of the Greeks. It was first incorporated into the Hellenistic kingdoms and was later integrated into the religious landscape of the Roman Empire. As Pleket has argued in his study of imperial mysteries, it is reasonable to suppose that ruler cult meant different things to different people. The well educated intelligentsia of the empire probably regarded it as little more than honorific in nature, but for many other inhabitants of lower social status it could have constituted a true form of religion. Ruler cult was in part a symbolic protocol that subjects of Greco-Roman antiquity employed to communicate with rulers. Emperor worship was not imposed from above upon unsuspecting subjects, but it was the product of a complex dialogue between the government and the governed. Its ubiquity and longevity resulted from the fact that it was based upon popular religious belief. As Pleket has argued, the evidence from the Greek East suggests that some of its inhabitants had no difficulty in using the symbolism of religious mysteries to express their notions of ruler worship. Similarly, the taurobolic inscription of L. Aemilius Carpus may suggest that such ideas were not foreign to inhabitants of the Latin west. In this regard it is noteworthy that the only two offices attributed to Carpus in this inscription are those of dendrophorus and sevir Augustalis, the former belonging to the cult of Magna Mater and the latter associated with the worship of the Roman emperors.

Of the other thirty-nine taurobolic inscriptions of imperial type, eight contain calendrical dates, nineteen do not, and the texts of the remaining twelve are lacunose or too fragmentary to allow us to determine whether or not they originally contained such information. Five of the eight documents that record specific dates are of interest in providing possible additional evidence for interaction between emperor worship and the Phrygian religious mysteries.[52] The first of these five documents comes from Corduba (*CIL* II. 5521 = *ILS* 4139) and records the performance of a taurobolium "ex iussu Matris deum" for the welfare of the empire on the Hilaria, March 25 in 238 A.D. Further evidence for a connection between the mystic and imperial rites of the Phrygian cult during the March ceremonies in Rome is to be found in Tertullian's *Apologia* 25.5, where the Christian polemicist ridicules the utter futility of pagan rites by conjuring up the image of the archigallus in Rome unwittingly shedding his own blood for the welfare of the dead Marcus Aurelius. The gibe makes sense when one is reminded that the emperor died on the Danubian frontier on March 17, 180 A.D., so that the archigallus could have been pictured as having performed his annual imperial taurobolium on March 24 in ignorance of the emperor's death.

The second of these five inscriptions, dating to the reign of Commodus, records the performance of an imperial taurobolium that extended over a four-day period:

> [. . . for the welfare of . . .][53] | . . . and of the imperial | house, and of the Colonia Copia Claudia Augusta of Lugdunum, | a taurobolium was performed by Q. Aquius Antoninia- | nus, pontiff for life, | in accordance with the prophecy of the archigallus Pusonius Ju- | lianus: begun twelve days before the kalends of May and com- | pleted nine days before the kalends of May in the consulship of L. Eggius Marullus | and Cn. Papirius Aelianus [i.e., 184 A.D.], the priest Aelius | C[astren]s[is] and the flute player Albius | Verinus presiding.[54]

The period of this celebration (April 20–23) included the Parilia of April 21, the day that the Romans celebrated as the birthday of Rome itself. The document's four-day time span corresponds to the four days of March 22–25 during which devotees of the Phrygian cult mourned Attis' death and resurrection. It therefore appears that the term 'taurobolium' in this document was regarded as not only pertaining specifically to the ceremony involving the sacrifice of a bull, but also was used to describe the mystic Phrygian rites as a whole. Moreover, this imperial taurobolium, it would seem, was deliberately observed so as to apply the symbolism of Attis' rebirth to the birthday of the Imperial capital.

Two other inscriptions from Lugdunum (*CIL* XIII. 1753–1754) are virtually identical in their wording to this text. XIII. 1754 records a four-day taurobolium for the welfare of Septimius Severus, observed May 4–7 of 197 A.D., whereas XIII. 1753 records a three-day imperial taurobolium for Septimius Severus and Clodius Albinus, celebrated May 9–11 in 194 A.D. At first glance, the choice of early May for the rites is puzzling, but a closer examination into the events of this period may offer an interesting solution. When Commodus was murdered on the last day of 192 A. D. (December 31), Pertinax succeeded him on the imperial throne on the very next day, January 1 of 193 A.D. Dio (LXXIII.10.3) informs us that Pertinax met his own death after a brief reign of 87 days, thus indicating that he died on March 28. His successor, Didius Julianus, enjoyed an even briefer reign of only 66 days (Dio LXXIII.17.5), meaning that he was killed on June 1. In the meantime, of course, Julianus' claim to the imperial throne was challenged both by Pescennius Niger, the governor of Syria, and by Septimius Severus, the governor of Upper Pannonia. Thanks to the *Feriale Duranum*, we know that Severus was proclaimed emperor by his army on April 9, which he later regarded as his *dies imperii.* After mobilizing support throughout the neighboring provinces, Severus proceeded to march upon Rome, but even before he reached the Imperial capital, the senate abandoned Didius Julianus and declared Severus emperor.[55] A lacunose entry in the *Feriale Duranum* has been plausibly restored to date this latter proclamation to May 21. Thus, the imperial taurobolia of early May recorded by *CIL* XIII. 1753–1754 might have been observed to mark the anniversary or anniversaries of important events of this period.

Septimius Severus, of course, eventually became Rome's next emperor and established the Severan dynasty, but he did so only by emerging victorious from four years of civil and foreign wars, first against Pescennius Niger in the East and then against Clodius Albinus in Gaul. Indeed, central to Severus' strategy in 193 was to mollify Clodius Albinus, the governor of Britain, with the title of Caesar while he dealt with much more immediate and serious threats to his imperial claim. According to Dio (LXXIII.15.1) Severus offered the junior imperial position of Caesar to Albinus before marching upon Rome, whereas Herodian (II.15.1–4) has Severus make the offer after Didius Julianus' death, after Severus' arrival in Rome, and at the beginning of the latter's campaign against Pescennius Niger. Faced with choosing between these two possible dates, we should probably side with Dio, because Herodian often treats events thematically rather than in strict chronological order. Thus, if Dio is to be believed, Severus invited Albinus to become his Caesar sometime between April 9 and June 1. We may therefore entertain the possibility that the dates recorded in these two taurobolic inscriptions relate to the *dies imperii* of Clodius Albinus and/or of his partnership with Septimius Severus. If so, these epigraphic texts offer us an important chronological datum for the momentous events of 193 A.D. According to this reconstruction, *CIL* XIII. 1753, dating to 194 and pertaining to both Septimius Severus and Clodius Albinus, would have marked the first anniversary of their joint rule, at a time when Severus was eager to have Albinus' full cooperation while he was engaged in a war with Pescennius Niger. According to the *Historia Augusta* (*Severus* 11.7), Clodius Albinus lost his life on February 19 of 197 while fighting in battle against Septimius Severus. Thus, *CIL* XIII. 1754, which honors Septimius Severus alone, would appear to have been conducted two and a half months after Albinus' death. Finally, the slight discrepancy in these two inscriptions' calendrical dates might also be significant. The imperial taurobolium performed in 194 obviously honored both Severus and Albinus, but a slightly different date was chosen three years later for the ritual so as to dissociate the two men from one another. April 9 might not have been firmly established as Severus' *dies imperii* until the following year.

The fifth taurobolic inscription of imperial type containing a calendrical date of possible relevance for imperial mysteries is inscribed upon a marble altar from the ancient town of Lactora in Aquitania (*CIL* XIII. 511 = *ILS* 4126). It records a taurobolium performed for the welfare of the Emperor Gordian, the entire imperial family, and for the state of Lactora itself. The ceremony was observed on December 8 of 241 A.D., one day before the end of the emperor's tribunician year, and two days before the beginning of a new one. In light of L. Aemilius Carpus' observance of an imperial taurobolium and mesonyctium on December 9/10, the proximity of this inscription's date to the emperor's tribunician day can hardly be accidental. Yet, it is difficult to say why the date of this taurobolium was not a day or two later; but if the text through epigraphic brevity has recorded simply the first day of a ceremony that lasted three or four days as those from Ludgunum just discussed, the taurobolic rites could have been completed on December 10 or 11.

Finally, another three-day imperial taurobolium may be attested in a fragmentary inscription from Lavinium dating to the year 212 A.D. (*AE* 1895 #120):

> . . . ius Maxi . . . | [sacerdos M(atris)] D(eum) M(agnae) I(daeae) L(aurentium) L(avinatium) et . . . | [sac]dotia taurob[olium] . . . | . . . it V, IIII, III . . . | . . . bres Aspro iteru[m et] | Aspro cos.

The first two and a half lines of this text seem to have contained the names of a priest and priestess of Magna Mater. The three descending Roman numerals in the fourth line can refer to nothing else except a calendrical date, and the "bres" in the fifth line must be the final syllable of a month's name (September, October, November, or December). In addition, since the dividing day (kalends, nones, ides) has not been preserved, there are, alas, nine different possibilities from which to choose:

Aug. 28–30	Oct. 3–5	Nov. 9–11
Sept. 9–11	Oct. 11–13	Nov. 27–29
Sept. 27–29	Oct. 28–30	Dec. 9–11

Since the three-day taurobolium recorded at Lugdunum for the year 194 A.D. is clearly of imperial type, we may conclude with some degree of confidence that this inscription also refers to an imperial taurobolium. Although choosing any one of the nine possible calendrical dates would be hazardous, it can nevertheless be observed that the latest of these dates (December 9–11) happens to coincide roughly with December 9/10, the emperor's tribunician day, and the day on which L. Aemilius Carpus observed his imperial taurobolium in Lugdunum in 160 A.D.

SOCIAL HISTORY OF THE TAUROBOLIUM

The document from Lactora discussed at the end of the previous section raises interesting questions concerning the social history of the taurobolium when it is compared with similar epigraphic texts from the same ancient site. Of the 39 inscriptions published in volume XIII of *CIL* from Lactora, 21 record taurobolia. In fact, this ancient site is second only to Rome itself in contributing the largest number of documents to the taurobolic corpus of inscriptions. As already mentioned, *CIL* XIII. 511 records the performance of a taurobolium for the welfare of the Emperor Gordian on December 8 of 241 A.D. Eight other inscriptions from Lactora (*CIL* XIII. 512–519) likewise record taurobolia performed by one man and seven women on the same day. Apart from the name of the celebrant, the wording of these eight texts is identical: "consecrated to the Mother of the gods, (person's name) received (accepit) the taurobolium with his/her own sacrificial victims, Traianius Nundinius being priest, six days before the ides of December in the consulship of our Lord Gordian for the second time and of Pompeianus." Since these inscriptions do not indicate that they were performed for the emperor's welfare, we may cautiously conclude that these eight taurobolia were personal and mystic in nature.[56] In contrast, according to the text of *CIL* XIII. 511 it was the town council of Lactora (ordo Lactoratium) who performed (fecit) the imperial taurobolium for Gordian's welfare. Like the other eight documents, this text records the name of the priest Traianius Nundinius in an ablative absolute; but unlike the other inscriptions, it includes another ablative absolute: "curantibus M. Erotio Festo et M. Carinio Caro." This latter clause should mean that M. Erotius Festus and M. Carinius Carus, members of the local town council, observed the imperial taurobolium on behalf of the entire body of decurions. Thus, on this one day in Lactora a total of ten individuals are known to have performed taurobolia.

The performance of such a large number of mystic and imperial taurobolia on the same day must have been the result of deliberate planning that involved both public magistrates and religious officials. Since the slaughter of ten bulls would have produced an enormous amount of fresh beef to be consumed, it can be reasonably supposed that local officials scheduled these private and public taurobolia on the same day, so that the religious needs of private individuals could be coordinated with the public cult in order to serve an important secondary function of providing the entire community with a banquet. Magna Mater's priest, Traianius Nundinius, would appear to have played a central role in organizing this event: for in addition to being the titular head of the public cult of the Phrygian goddess in Lactora, he was the one who presided over the administration of the private rites of the same cult. By including the phrase "with his/her own sacrificial victims (hostiis suis)" in the eight mystic taurobolic inscriptions the private celebrants were probably advertising their own euergetism in contrast to the two imperial taurobolia, whose animals were doubtless paid for with municipal funds. Consequently, this series of documents allows us to see how the cultic activities of single individuals might not simply serve personal religious goals, but by being subsumed within the public cult they could also constitute a civic benefaction.

Several Other epigraphic texts record similar multiple taurobolia, three of which openly stress their performance at the expense of the celebrants. The one document that advertises its euergetism more than any other is *CIL* XII. 11,567 from Dea Augusta among the Vocontii of Narbonensis, recording three taurobolia for the welfare of the Emperor Philip, his son, and the imperial consort. The text begins with the impressive proclamation: "a sacrifice of three bulls with their own victims and all the accouterments was made to the Great Idaean Mother of the gods by L. Dagidius Marius, . . .

Verullia Martina, and their daughter Verullia Maria." Two other texts more briefly record multiple taurobolia undertaken at the expense of the worshippers. *CIL* XII. 1 from the Maritime Alps reads: "Valeria Marciana, Valeria Carmosyne, and the priest Cassius Paternus celebrated the taurobolium to the Idaean Mother at their own expense." Likewise, *CIL* XII. 1744 simply says: "the dendrophori of Valentia performed the taurobolium to the Great Idaean Mother of the gods with their own money." As in the case of the rites performed at Lactora on December 8 of 241 A.D., we are probably justified in assuming that in these instances as well the flesh of the sacrificed animals was consumed in some kind of large scale banquet—Hence, the celebrants' care in recording their expenditures.

Four inscriptions offer interesting glimpses into the mechanisms of official organization and promotion of taurobolia by suggesting that they were sometimes observed by or on behalf of an entire community or province. According to one text from Dea Augusta (*AE* 1889 #81) "the *res publica* of the Vocontii performed the taurobolium" for the welfare of Septimius Severus and his two sons and wife. The nature of this celebration might have been in part similar to the one in Lactora in which the town council (ordo Lactoratium) is recorded as the celebrant, but two persons actually observed imperial taurobolia on behalf of the entire body of decurions. One document from Narbo reads: "with collected contributions | the Narbonensians publicly celebrated | for the Mother | of the gods | a taurobolium proclaimed | by her order."[57] The wording of the text suggests that the community enjoyed a public banquet that included beef supplied by several taurobolic sacrifices financed by private donations. Unfortunately, the text does not specify which persons or how many people actually observed a taurobolium, and how they were selected. Moreover, the celebration was enjoined upon Narbo by the goddess herself, which probably means that the taurobolium was prompted by some local portent or dream experienced by a person of note in the community.

Two other inscriptions from Narbo mention a taurobolium with respect to the entire province. *CIL* XII. 4329 merely consists of the two words "tauropolium provinciae," inscribed upon a marble commemorative altar. The other Narbonese text is only slightly more informative:

> On the command of the Mother of the gods a tauro- | polium of the province | of Narbonensis | was performed through C. Batonius | Primus, flamen Augustalis, | for the welfare of the Lord | Emperors L. Septimius Severus | Pius Pertinax Augustus Ara- | bicus Adiabenicus Parthi- | cus Maximus, M. Aurelius | Antoninus Augustus . . .[58]

The flamen Augustalis mentioned here was the titular head of the cult of the Roman emperors for the entire province of Gallia Narbonensis. A literal interpretation of the document's wording would seem to indicate that this religious official, not otherwise known to be associated with the cult of Magna Mater, personally performed an imperial taurobolium on behalf of the entire province, and that the ceremony might have been commanded by the goddess in a dream, possibly experienced by Batonius Primus himself. If Batonius alone observed an imperial taurobolium for the entire province, his action would appear to resemble the yearly imperial taurobolium experienced by the archigallus in Rome mentioned by Tertullian.

Three taurobolic inscriptions of imperial type indicate that the ceremonies were performed "in accordance with the prophecy of the archigallus (ex vaticinatione archigalli)."[59] The latter phrase strongly suggests that the archigalli in the local Phrygian cults could exercise considerable influence with respect to celebrating taurobolic rites for the emperor's welfare. The word 'prophecy' may mean that on occasions archigalli were inspired to make formal pronouncements in the form of a conditional statement such as "if the community or so and so undertakes to observe a taurobolium, then all will go well with respect to . . ." The legal significance of the phrase "ex vaticinatione Archigalli" is suggested by s.148 of the *Fragmenta Vaticana* concerning exemptions from guardianship. According to this legal compilation of imperial constitutions dating to the late fourth century, "he who offers sacrifice in Portus for the emperor's welfare in accordance with the prophecy of the archigallus is also excused from guardianship."[60] The Portus mentioned here is, of course, Portus Augusti, the harbor community at the mouth of the Tiber, established by the Emperor Claudius. It therefore appears that the Imperial government (and perhaps municipal governments as well) granted legal exemptions from guardianship in return for the performance of taurobolic rites that benefited one's community. Thus, this terse legal statement reinforces the foregoing interpretation of imperial taurobolia as representing in part a form of public service to one's community by contributing to the celebration of a public banquet. Consequently, we must consider the possibility that many of the epigraphic texts recording the performance of a taurobolium may not have been primarily or exclusively intended to advertise one's personal religiosity or to convey specific religious information to posterity, but rather they may have been designed to publicize one's civic minded behavior, which could be of value in making one eligible for a legal exemption. Hence, the data contained in many taurobolic documents are not only valuable for the history of a religious cult, but they also demonstrate how the cult of Magna Mater was embedded in the political, social, and legal institutions and practices of the Roman Empire.

TAUROBOLIC REPETITIONS, BIRTHDAYS, AND OTHER DATES

Of the seventy-five taurobolic inscriptions that appear to be of mystic type in that they do not contain the formula *pro salute imperatoris*, the texts of thirty-three contain calendrical dates, twenty-six do not, the texts of twelve are lacunose or too fragmentary to allow us to determine whether or not they included such information, and four refer to the celebrant's birthday without recording an actual calendrical date. Thus, roughly half (33+4) of these documents directly or indirectly mention a specific day of the year. An examination of these dates reveals that many had symbolic significance for the meaning of the taurobolium and/or for the person who observed the ceremony.

A substantial body of data demonstrates that from very early times the Romans were in the habit of recording and commemorating the dates and anniversaries of major events. In fact, much of the Roman religious calendar was nothing more than cultic commemorations of the dedication days of temples. Likewise, the calendrical dates recorded in the *Fasti Triumphales* exhibit a similar phenomenon in the sphere of military affairs. The recording of these religious and military dates was doubtless the duty of the pontiffs. This practice stemmed from their belief that each day in the year possessed its own characteristics that could and, in the interest of the Roman state's welfare, should be divined empirically from direct observation. Similarly, literary sources indicate that at least from the end of the second century B.C. onwards Romans made careful note of their day of birth and celebrated it.[61]

Not only was Augustus' birthday widely celebrated throughout the empire (e.g. *ILS* 112), but the province of Asia changed its calendar so that this day would be the first day of the year (*OGIS* 458). The ubiquity of the Roman concept of the birthday during Imperial times is well illustrated by two documents from Spain and one from Britain. The former (*AE* 1967 #229–230) record dedications to Jupiter Optimus Maximus for the welfare of Antoninus Pius and Commodus respectively on the birthday of the military standard of Legio VII Gemina, i.e., the anniversary of the day on which the standard was officially consecrated and received by the legion. The latter document, discovered at Vindolanda (Bowman and Thomas 1994 256–259), is a letter written on a very thin piece of wood, in which a woman invites her sister to her birthday party.

The close association between a Roman's identity and his or her birthday is quite apparent from both literary and non-literary evidence. The Historia Augusta, for example, is careful to note that Caligula and Commodus shared the same birthday, as did Nero and L. Verus; and the imperial biographies thereby imply that the depravity and failings of Commodus and L. Verus were in part due to this coincidence.[62] Numerous inscriptions show that the deceased stipulated in their wills to have their birthday memorialized in various ways.[63] Thus, when we encounter four taurobolic inscriptions that refer to a person's birthday, we cannot doubt but that the celebrants deliberately chose to observe the ceremony on this day for its obvious symbolic significance in representing the person's biological birth and thus in helping to guarantee their spiritual rebirth after death. Two of the four "birthday" taurobolia come from Lusitania and read as follows:

> Consecrated to the Mother of the gods, | the two Irinaei, father and | son, underwent the criobolium | on their birthday, the priests being | L. Antistius Avitus | and G. Antistius Felicis- | simus.[64]

> Consecrated to the Mother of the gods, | Valeria Avita | gave and dedicated | the altar of the taurobolium | of her returned birthday, | the priest being Doccyricus Vale- | rianus and the archigallus | being Publicius Mysticus.[65]

The third of these inscriptions is to be found on a taurobolic altar from Bordeaux and simply reads: "Valeria Jullina and Julia Sancta to the *vires* of their birthday."[66] As already noted, it was standard practice to bury the testicles of the sacrificed bull beneath the consecrated taurobolic altar. Given the simple nature of this inscription, we should not interpret *natalici* metaphorically as referring to the women's spiritual awakening and birth. Rather, the words *natalici viribus* should be taken in their literal, mundane sense, thus indicating that the taurobolium was observed on the women's birthday.

The fourth "birthday" taurobolic inscription occurs on a large limestone altar discovered at Metz on the Moselle in northeastern France (*CIL* XIII. 11,352) and contains a consular date corresponding to the year 199 A.D. Unfortunately, the text cannot be read with certainty, but it clearly contains the two phrases *ob natalicium* and *ex iussu* and seems to have recorded the refurbishing of the altar. Thus, we may cautiously conclude that the devotee, commanded by Magna Mater in a dream or by some other means, took some measures in repairing his or her taurobolic altar in connection with his or her birthday. Thus, these four short inscriptions strongly suggest that like many other Romans, some of Cybele's worshippers attached considerable personal and religious significance to their own birthdays and even went so far as to have the anniversary of their taurobolium coincide with it.

The following list sets forth the 33 taurobolic inscriptions of presumed mystic type that contain calendrical dates, arranged in order of these dates.

Feb. 26, 295: *CIL* VI. 505.
Mar. 12, 377: *CIL* VI. 511.
Apr. 5, 383: *CIL* VI. 501.
Apr. 5, 383: *CIL* VI. 502.
Apr. 9, 228: *CIL* IX. 1538.
Apr. 14, 305: *CIL* VI. 497.
Apr. 15, 313: *CIL* VI. 507.
Apr. 19, 319: *CIL* VI. 508.
Apr. 29, 350: *CIL* VI. 498.
May 15, 199: *CIL* XIV. 39.
May 23, 390: *CIL* VI. 503.
May 23, 390: *CIL* VI. 512.
May 27, 387: *IG* III. 173.
June 16, 370: *CIL* VI. 509.
July 19, 374: *CIL* VI. 499.
July 19, 374: *AE* 1953 #238.
July 22, 228: *CIL* IX. 1542.

Aug. 13, 376: *CIL* VI. 504.
Aug. 13, 376: *CIL* VI. 510.
Oct. 7, 134: *CIL* X. 1596.
Oct. 18, 176: *CIL* XIII. 505.
Oct. 18, 176: *CIL* XIII. 506.
Oct. 18, 176: *CIL* XIII. 507.
Nov. 20, 186: *CIL* X. 4726.
Nov. 26, 235: *CIL* IX. 3015.
Dec. 8, 241: CIL XIII. 512.
Dec. 8, 241: *CIL* XIII. 513.
Dec. 8, 241: *CIL* XIII. 514.
Dec. 8, 241: *CIL* XIII. 515.
Dec. 8, 241: *CIL* XIII. 516.
Dec. 8, 241: *CIL* XIII. 517.
Dec. 8, 241: *CIL* XIII. 518.
Dec. 8, 241: *CIL* XIII. 519.

As will be argued below in connection with taurobolic repetitions, the three inscriptions dating to October 18 of 176 A.D. are best understood in terms of a public celebration at Lactora in honor of Commodus' decennalia as Caesar. Thus, these documents should actually be set aside along with the other eight inscriptions from Lactora, dating to December 8 of 241, which have already been discussed in terms of another public celebration commemorating the emperor's tribunician day. We are therefore left with 22 dated texts.

Some obvious patterns are immediately discernible. Relatively few taurobolia occur during the seven months from September to March, whereas April and May alone account for half of the documents (11 of 22). April 4–10, of course, witnessed the celebration of the Megalensian Games in honor of Magna Mater, but the month also included the *Ludi Ceriales* of April 12–19 and the *Ludi Florales* that ran from April 28 to May 3. Thus, virtually the entire month of April, which accounts for seven of the inscriptions with calendrical dates, was dominated by festivals surrounding female divinities of fertility. Given the syncretic tendencies of Paganism during Imperial times, it is reasonable to suppose that many of Magna Mater's devotees regarded both Ceres and Flora as manifestations of the Great Mother herself. In fact, the inscription dating to April 19 of 319 A.D. further defines the calendrical date with the term *Cerialibus*, thus stressing the connection between the taurobolium of Magna Mater and the cultic birthday of Aventine Ceres. In this regard it is noteworthy that this is the only inscription in the entire taurobolic corpus that uses the name of a Roman festival to indicate the date.

Perhaps the most surprising absence in the list of calendrical dates pertains to the period of March 15–27. Nevertheless, it is quite likely that these days were held to be of such great religious solemnity that they were reserved exclusively for commemorating the birth, life, death, and resurrection of Attis; whereas the initiation of neophytes was scheduled for the more festive days in April. One explanation for the heavy concentration of taurobolia in the two months of April and May can perhaps be found in the widespread ancient belief in astrology: for the zodiacal signs of these two months were Aries (the Ram) and Taurus (the Bull) respectively,[67] which worshippers doubtless associated with the ceremonies of the criobolium and taurobolium that involved the sacrifice of a ram and bull. Thus, it appears likely that according to ancient belief criobolia and taurobolia were most efficacious when these two astrological signs were dominant in heaven and held sway over human affairs.[68]

Other, more personal reasons may account for the calendrical dates of some taurobolia. It is certainly possible that some of the other eleven taurobolia dating to months other than April and May could have been observed on the celebrant's birthday; and this important fact, though well known to the initiate, has been lost to us through the laconic and formulaic language of Latin epigraphy. A second explanation could be the devotee's close association with some other aspect of pagan cult. This seems to have been the case with respect to *CIL* VI. 504 and 510, both dating to Aug. 13 of 376 A.D. and being the inscriptions from the taurobolic altars respectively of Ulpius Egnatius Faventinus and Sextilius Agesilaus Aedesius: for in the list of their priestly offices set forth in these texts we encounter the title 'Hierophant of Hecate'. August 13 was the *dies natalis* of the famous temple of Diana on the Aventine, whose construction and dedication were attributed by Roman tradition to King Servius Tullius.[69] Moreover, since the Greek divinity Hecate was generally equated with Roman Diana, these two Roman senators obviously chose to undergo Magna Mater's rites of initiation on a day that was associated with another prominent goddess to whom the initiates were likewise devoted. July 19, the day on which two other Roman senators observed taurobolia in 374 A.D., marked the rising of Sirius, the Dog Star in the constellation Orion, which signalled the beginning of the Nile's inundation. Nevertheless, there is nothing to indicate in the religious offices held by these two senators that they were worshippers of Isis and Serapis.

Three epigraphic texts indicate that the taurobolium commemorated by the inscription was a repetition of an earlier observance of the ceremony. One of these (*CIL* X. 1596) is the earliest of the metroac taurobolic inscriptions, dating to 134 A.D., recording that Herennia Fortunata repeated (iterata est) an *ecitium taurobolium* of Venus Caelestis and a *pantelium*. The other two date to the late fourth century. *CIL* VI. 502 (= *ILS* 4150), dating to April 5 of 383 A.D., records that a woman of senatorial rank and a *sacerdus* (sic) *maxima* of the Great Mother repeated both the taurobolium and criobolium (taurobolio criobolioque repetito). The third text (*CIL* VI. 512 = *ILS* 4154), dating to May 23 of 390 A.D., indicates that a Roman senator repeated the taurobolium after the completion of twenty years (iterato viginti annis expletis). Similarly, in a fourth document dating to Aug. 13 of 376 A.D. (CIL VI. 504 = *ILS* 4153), another Roman senator appends to the standard taurobolic inscription two lines of dactylic hexameter in which he vows to repeat the ceremony twenty years hence: "vota Faventinus bis deni suscipit orbis, | ut mactet repetens aurata fronte bicorne." It is therefore apparent that repetition of the taurobolium was an established practice throughout the entire recorded history of the rite, and that during the closing decades of paganism it was not uncommon for devotees to repeat the ceremony after an interval of twenty years.

The repetition of the taurobolium harmonizes with other evidence concerning the ancient mystery religions. Lucius in Apuleius' *Metamorphosis* XI.26–27, after being initiated into the cult of Isis, is commanded by the goddess in a dream to undergo other initiatory rites in order to become a devotee of Serapis. In this regard it is noteworthy that according to *CIL* X. 1596 Herennia Fortunata repeated the taurobolium "by command of the goddess (inperio deae)." Since the ancient mystery religions were served by many different functionaries who were probably organized hierarchically, advancement from one office to another is likely to have involved the observance of special rites. Modern scholars generally suppose such a scheme for the seven grades of Mithraism. In the cult of Magna Mater these stages of advancement could have required in some instances the performance of a taurobolium, which the worshipper may have already observed once before. In this context may belong an inscription from Beneventum (*CIL* IX. 1541) recording a taurobolium observed by a priestess of the second order that was supervised by a priestess of the first order. Alternatively, repetition of the taurobolium could have resulted from a personal decision to rededicate oneself to the cult or to symbolize and express the intensification of one's religious devotion.

Simply on the basis of two inscriptions belonging to the late fourth century of our era we should not deduce that the taurobolium was generally believed to be efficacious for only twenty years, after which it needed to be repeated.[70] Rather, given the syncretic and henotheistic tendencies of pagan religious thought during late antiquity that transformed deities such as Magna Mater and Attis into supreme cosmocratoric divinities, it seems better to surmise that just as the god of the Christians from ancient times to the present day has often been viewed by devout worshippers not only as the sole agent of their personal spiritual salvation but also as the omniscient and omnipotent divine architect of health and prosperity in their daily lives, the Phrygian deities were viewed in similar terms by their pagan devotees, who in accordance with the habits of ancient votive religion expressed their gratitude to the two great gods for the overall happiness in their lives by the periodic repetition of the taurobolium. Once again, such an ancient attitude is clearly evident in the eleventh book of Apuleius' *Metamorphosis*: for in addition to serving as the divine agent for Lucius' own spiritual salvation, Isis alone of all the gods intervenes to rescue Lucius from his asinine form and hardships, and at the very close of the book his piety and thankful subservience to the goddess receive their due reward in the form of Lucius' success as an advocate in the Roman law courts.

One modern scholar (Moore 1924) has attempted to explain the twenty-year taurobolic interval as representing half of a generation of forty years and as corresponding to the ephebic attainment of manhood at the age of twenty, well attested in the Greek East. The ephebic connection is definitely attractive since, as already discussed, the four earliest non-metroac taurobolic and criobolic inscriptions come from Asia Minor and concern rodeo-like public spectacles in which ephebs participated in activities involving the chasing and overpowering of bulls and rams. Yet, even though this may account for the ultimate origin of the twenty-year interval, another factor is likely to have played an important secondary role in its acceptance into the Phrygian cult during the later Roman empire. At least since the time of Hadrian onwards, whenever an emperor came to power, public vows were made to commemorate the emperor's completion of ten years of rule by the celebration of *ludi votivi decennales*; and at the time of the first decennalia vows for a second imperial decennium were probably offered.[71] As a chronological list of Roman emperors makes clear, several rulers reign long enough to enjoy their decennalia, but relatively few were fortunate enough to celebrate a vicennalia, their twentieth imperial anniversary; and from the second century onwards Constantine alone ruled long enough to celebrate a tricennalia. The degree to which the notion of imperial decennalia, vicennalia, and tricennalia was prevalent in popular Roman culture can perhaps be detected from *ILS* 4937, an inscription from a statue base set up in honor of Coelia Claudiana, chief Vestal virgin of the late third century of our era: for not only does the dedication commemorate her vicennalia as chief priestess, but it also expresses the hope that she will complete an additional ten years in order to enjoy her tricennalia (sic XX, sic XXX feliciter!).

Three taurobolic inscriptions from Lactora (*CIL* XIII. 505–507) suggest that already as early as the second century of our era taurobolia were performed to celebrate imperial decennalia. The ceremonies recorded by these three texts were observed on October 18 of 176 A.D. Two other taurobolic altars from Lactora (*CIL* XIII. 508–509), as is evident from the name of the presiding priest, belong approximately to the same period of time and could even date to the same day as the other three altars. In view of the taurobolic rites observed at Lactora on December 8 of 241, in which two imperial and eight mystic taurobolia were performed in honor of the emperor's tribunician day, we are justified in suspecting that these three earlier mystic taurobolia were likewise celebrated together with imperial ones in honor of some major event relevant to the imperial family.[72] On October 12 of 166 A.D. Marcus Aurelius conferred the title of Caesar upon the five-year old Commodus (*Hist. Aug. Marci* 12.8 and *Commodi* 11.13). Thus, the taurobolic rites celebrated at Lactora in 176 A.D. could have been intended to commemorate Commodus' decennalia as Caesar. This conjecture is rendered plausible by the great importance given to this anniversary by the revolt of Avidius Cassius in Syria during the spring of the preceding year and Marcus' concomitant rapid advancement of Commodus as his chosen successor. The boy's coming of age was marked by his assumption of the toga virilis on July 7 of 175; then after he and his father returned to Rome in the autumn of 176, they were jointly hailed as *imperatores* on November 27, Commodus celebrated a triumph with his father on December 23, and the two entered upon the consulship together on January 1 of 177.[73] Thus, this interpretation of taurobolic dedications from Lactora raises the possibility that the celebration of imperial decennalia and vicennalia influenced the practice of repeating the taurobolium after a similar interval of time.

6 The Non-Christian Origin of Christmas[1]

Christianity is the single most important religious legacy passed on by the ancient Roman world to later times. Out of Christianity's complex set of beliefs and practices, Christmas, the celebration of Jesus' birth on December 25, is the most striking example of how a temporal concept with strong religious associations was taken from Roman paganism, absorbed into Christianity, and given a new meaning. Long before it became Christmas, December 25 was simply the day of the winter solstice; and before being associated with Jesus' birth, the day was the birthday of the Unconquered Sun (Sol Invictus) and of Mithras, the central figure of a mystery religion that was very popular during Imperial times. The process by which December 25 went from being the winter solstice to the birthday of Jesus is a fascinating story and reveals much about the religious history of the Roman Empire and how Christianity appropriated unto itself aspects of paganism, adapted them, and was thereby better equipped to quash all its pagan rivals and to absorb their worshippers into its own following.

WHEN WAS JESUS BORN?

Determining the date of Jesus' birth has been a major historical problem since at least the second century of our era. *The Gospel According to Mark*, the earliest of the four gospel narratives to be written, offers nothing at all concerning Jesus' life before his ministry.[2] *Matthew* and *Luke*, written after *Mark* and probably attempting to satisfy the natural curiosity of potential converts, filled this void by prefacing their accounts of Jesus' ministry with two very different nativity narratives, which were carefully crafted to prove that Jesus had been born the son of God and was the true Messiah.[3] According to Matthew (2) Joseph and Mary were living in Bethlehem in Judea during the reign of King Herod the Great when Jesus was born. His birth was heralded by a wondrous star that brought to Jerusalem Persian Magi inquiring after the birth of the King of the Jews. Following the Magi's visit to Bethlehem to the home of Joseph and Mary to present costly gifts to the newborn king, Joseph, warned by an angel in a dream, took Mary and Jesus and fled into Egypt to avoid King Herod's slaughter of infants that was intended to kill the baby Messiah. When Herod died shortly thereafter, Jesus' family returned to Judea, but since Herod's son Archelaus was ruling in his father's place, they moved north into Galilee and settled at Nazareth. *Luke* (2.1–20), on the other hand, places Joseph and Mary's original abode at Nazareth and describes how they were obliged to travel to Bethlehem in order to be assessed in a census when Quirinius was governor of Syria. While staying in Bethlehem, Mary gave birth to Jesus; and in response to an angel's message concerning the birth of a savior, shepherds from the surrounding area came into Bethlehem and marveled at the newborn babe lying in a manger as foretold by the angel.

In order to have Jesus, a Galilean from Nazareth, fulfill the expectation that the Jewish Messiah would be born in Bethlehem, the town of King David, *Matthew* and *Luke* set forth two divergent explanations. *Matthew* describes Bethlehem as the original home of Joseph and Mary and has them move to Nazareth on their return from Egypt. *Luke*, on the other hand, regards Nazareth as Joseph and Mary's original domicile and has Jesus born in Bethlehem when his parents traveled thither during the time of a Roman census. Moreover, the two Gospels seem to disagree as to when Jesus was born. *Matthew*, on the one hand, clearly dates Jesus' birth to the reign of King Herod the Great. Thanks to Josephus' mention of a lunar eclipse and of the celebration of Passover (*Ant. Iud.* XVII.6.4 and 8.1–9.3 with Kidger 1999 46–49), we know that Herod died sometime between March 13 (the eclipse) and April 10 (Passover) of 4 B.C. *Luke*, on the other hand, dates Jesus' birth to the time when Quirinius, the governor of Syria, conducted a census of Judea. Josephus' detailed account of Jewish affairs for this period is quite clear in indicating that P. Sulpicius Quirinius was sent out to govern Syria in 6 A.D., after Augustus had removed Archelaus, King Herod's son, from his rule over Judea and had the area annexed to the Roman province of Syria (*Ant. Iud.* XVII.13.5, XVIII.1.1–2, and 2.1). Thus, there appears to be a chronological discrepancy of about a decade between *Matthew* and *Luke* concerning the year in which Jesus was born; and as a result, the matter has attracted considerable modern scholarly attention, largely because upon it hinges *Luke's* credibility as a biographer of Jesus' life.[4]

According to Josephus (*Ant. Iud.* XVII.8–XVIII.1), when Herod died, Augustus appointed Herod's son Archelaus to replace his father as Rome's client king to rule Judea, but when Archelaus had proven himself incapable of governing the Jews, Augustus removed him from his position, abolished Judea's status as a client kingdom, annexed the area to the Roman province of Syria, and placed Judea under the administration of a Roman equestrian prefect, who was subordinate to the Roman provincial governor of Syria. In keeping with this major change in Judea's status within the Empire, the Romans for the first time conducted a census of the inhabitants of Judea for the purpose of imposing Roman taxation upon these new provincial subjects. The census, however, encountered resistance led by Judas of Galilee; and this opposition formed the beginning of the Zealot movement that increased in strength over the next sixty years, finally culminating in the Jewish Revolt in 66 A.D. (see Josephus *Ant. Iud.* XVIII.1, XX.5.2, *Bell. Iud.* II.7.1, and VII.8.1). This narrative of events is coherent, logically consistent, and agrees with what else we know about the workings of Roman Imperial administration. Josephus' account has therefore been generally accepted as historically accurate by modern historians of the Roman Empire; but since it seems to contradict *Luke's* account of Jesus' birth, modern scholars, especially of the *New Testament*, have developed a number of arguments in attempting to reconcile the discrepancies among *Matthew, Luke*, and Josephus.[5]

In order to assign Quirinius' census of Judea to the reign of King Herod, some scholars have equated the census of *Luke* 2.1–2 with Augustus' second census of 8. B.C. recorded in Chapter 8 of his *Res Gestae*. This census, however, along with the other two mentioned by Augustus, was not a census of all inhabitants of the Roman Empire, as *Luke* says, but it most likely involved only the Roman inhabitants of Italy. Taking a census in the provinces was an entirely different matter; and as long as Judea was a client kingdom of Rome under the rule of Herod and Archelaus, the Romans would not have carried out a systematic census of all of Judea's inhabitants. Nevertheless, scholars desirous of harmonizing *Matthew, Luke*, and Josephus have further suggested that Quirinius might have been governor of Syria on two different occasions, that Josephus reports only his second governorship beginning in 6 A.D., and that Quirinius could have been governor of Syria for the first time during the later years of Herod's reign. Indeed, in his commentary on Augustus' *Res Gestae* Mommsen (1883 161–78) devoted considerable space to reconstructing the career of P. Sulpicius Quirinius, in which he argued in favor of him having held two Syrian governorships by restoring the latter's name in an acephalous Tiburtine *elogium* (*ILS* 918). This inscription, discovered near Tibur in 1764 and now preserved in the Vatican's Galleria Lapidaria, records the career of an eminent Roman of the early principate; but since the top part of the stone has been broken off, all that remains is the mention of an unknown king and the recapturing of an unspecified area for the Roman people, for which the man was awarded *ornamenta triumphalia* along with the senate decreeing two *supplicationes*. The inscription then ends by listing the man's proconsular governorship of Asia and a second command as *legatus pro praetore Divi Augusti*, in which capacity he governed Syria and Phoenicia.[6] Mommsen suggested that the war mentioned at the beginning of the damaged inscription refers to the one known to have been waged by Quirinius against the Homonadeis, a people who dwelled in the mountains of Pamphylia. The war is mentioned briefly by Tacitus (*Ann.* III.48) and Strabo (XII.6.5); and its date is approximately determined by the erection of Roman milestones in the region dating to late 6 B.C. (*ILS* 5828).

Ramsay, eager to vindicate *Luke's* veracity, gladly accepted Mommsen's reconstruction of Quirinius' public career and developed it even further by offering his own detailed reconstruction of his war against the Homonadeis.

According to Ramsay (1917 236–239), immediately following his consulship of 12 B.C., Quirinius was sent out to Syria; and over the course of the next few years conducted the war, whose end Ramsay dated from the milestones erected in late 6 B.C. One major obstacle to this reconstruction of events is that Josephus' narrative seems to supply us with an adequate number of Roman governors of Syria during the last years of Herod's reign: M. Titius c.12 B.C., C. Sentius Saturninus c.8 B.C., and P. Quinctilius Varus c.6–4 B.C. Since provincial governors under Augustus held their commands for at least a minimum of three years, it is difficult to fit a Syrian governorship of Quirinius somewhere in these years. Ramsay (1917 271–272) circumvented the problem by suggesting that while Quirinius was busy waging war against the Homonadeis, a legate, such as C. Sentius Saturninus, served as Quirinius' deputy in administering the affairs of Syria, including the conduct of the census; but as Taylor (1933 124) pointed out, there is no evidence for provincial commands being divided in this way between two legates. Modern Roman historians also subsequently undermined Ramsay's reconstruction by arguing persuasively that the most logical base from which to have launched the Homonadensian War was not from the east from Syria, but from the north from Galatia, so that it seems far more probable that Quirinius was governor of Galatia when he waged his war against the Homonadeis.[7] Furthermore, the acephalous *elogium* from Tibur does not state that the unknown noble was twice legate of Syria, but as *legatus Augusti pro praetore* for a second time he received the province of Syria and Phoenicia to govern. Since Tacitus (*Ann.* III.48) informs us that Quirinius came from Lanuvium, not Tibur, assigning *ILS* 918 to his career is far from certain, especially since Groag, Taylor, and Syme have developed strong arguments for relating the inscription to M. Plautius Silvanus, M. Titius, or L. Calpurnius Piso respectively.[8] Thus, in conclusion, the scholarly effort to salvage *Luke's* credibility by assigning an additional Syrian governorship to Quirinius appears to have no historical validity and should be abandoned.

When standard historical critical analysis is applied to the rest of the two nativity narratives, all their principal features likewise do not hold up well to careful scrutiny. Modern Biblical exegesis renders it likely that much of the two accounts is unhistorical and in part has been patterned after incidents from the *Old Testament* in order to portray Jesus as the true Messiah. For example, King Herod's slaughter of the infants in *Matthew* probably never occurred. It is most likely a retelling of Pharaoh's attempt to kill all Jewish newborn males in the first chapter of *Exodus*. The apparent purpose behind the Matthean story was to establish a clear connection between Moses and Jesus, which is further underscored by the flight of the latter's family into Egypt and return therefrom. *Matthew* thereby depicts Jesus as Moses' successor. In fact, *Matthew* quotes *Hosea* 11.1 to make it appear that by descending into and returning from Egypt Jesus fulfilled an important prophecy; but when one consults the text of *Hosea*, the alleged prophecy turns out to be nothing more than an allusion to the early Hebrews' exodus out of Egypt, not a prediction of some future event. *Matthew's* miraculous star that guided the Magi from the East to Jerusalem and thence to the very house of Joseph and Mary in Bethlehem is best understood not as some historical astronomical event such as Halley's Comet, a supernova, a meteor, or a rare conjunction of planets (contra Hughes 1979 and Kidger 1999), but as the Matthean reworking of an episode in *Numbers* 22–24 involving a Moabite wise man seeing in a dream a star coming out of Jacob to signify the Hebrews' rise to dominance in Canaan (Brown 1993 190–196). According to this story, when the Hebrews under Moses arrived in Moab, the local king summoned a wise man and diviner named Balaam to curse the newcomers, but instead, following the visitations of angels and speaking with God himself in dreams, Balaam blessed the Hebrews and said that he saw a star coming out of Jacob and a scepter that would smite Moab.

Likewise, the material contained in the first two chapters of *Luke* appears to be largely, if not entirely, unhistorical. For example, the unexpected birth of John the Baptist to the aged Elisabeth and Zacharias (*Luke* 1.5ff) is clearly patterned after Isaac's birth to the aged Abraham and Sarah in *Genesis* 17.15ff. Moreover, of the four Gospel narratives, *Luke* 1.36 alone contains the assertion that Mary and Elisabeth, the mothers of Jesus and John the Baptist, were cousins. The claim is far too neat to be accepted as credible, because the whole purpose of *Luke's* first chapter is to coordinate the births of Jesus and John into a single divine plan and to subordinate John to Jesus even while the former was in his mother's womb; whereas all four Gospels (including *Luke*) suggest that there was rivalry between Jesus and John and their followers, which the early Christians tendentiously reshaped as John's subordination to Jesus.[9] *Luke*, however, at 3.1–2 does provide us with what appears to be a solid chronological fact. After filling two chapters with stories concerning the births of Jesus and John the Baptist, *Luke* begins its narrative of the two men's ministry with a very careful dating formula: "In the fifteenth year of the reign of Tiberius Caesar, while Pontius Pilate was governing Judea, while Herod was tetrarch of Galilee, while his brother Philip was tetrarch of Ituraea and Trachonitis, and while Lysanias was tetrarch of Abilene, 2 when Annas and Caiaphas were high priests, the word of God came upon John, the son of Zacharias, in the wilderness." Then after briefly summarizing John's message and describing his baptism of Jesus, Luke (3.23) further states, "and at the outset (*archomenos*) Jesus himself was about (*hosei*) thirty years old."

When Augustus died on August 19 of 14 A.D. (Suet. *Aug.* 110.1), Tiberius unofficially assumed power as his successor (Tac. *Ann.* I.5); but his official reign as emperor did not actually begin until nearly a month later on September 17.[10] Thus, we may date the beginning of Tiberius' principate to the late summer of 14 A.D., so that the fifteenth year of his reign would have run from late summer of 28 to late summer of 29 A.D. Nevertheless, *Luke's* attempt to be quite precise in dating the beginning of Jesus' ministry lacks precision, because the dating formula applies to the beginning of the ministry of John the Baptist and fails to inform us how much time elapsed from its beginning to John's baptism of Jesus. In addition, what does Luke actually mean by characterizing Jesus as being "about thirty years of age?" Could he have been as young as 27 or as old as 34? Thus, even if *Luke* were correct in every respect in 3.1–2 and 3.23, too many uncertainties still exist to allow us to establish unequivocally the year in which Jesus was born; but even when we make allowances for *Luke's* chronological imprecision, it is very likely that Jesus' birth did not occur during the reign of King Herod, but some time shortly after Herod's death. If so, then neither *Matthew* nor *Luke* succeeded in fixing the exact year in which Jesus was born. *Matthew's* nativity narrative is clearly interested in portraying Jesus as the New Moses, and this is largely accomplished through the tale of King Herod's slaughter of the infants, but since it is patterned after the opening verses of *Exodus*, the tale should not be given any historical credence. Similarly, as already demonstrated, *Luke's* dating of Jesus' birth to the census of Judea while Quirinius was governor of Syria beginning in 6 A.D. fails to hold up to historical scrutiny. It therefore appears likely that when *Matthew* and *Luke* were written about a half a century after the crucifixion, there was no reliable information still available concerning the circumstances surrounding Jesus' birth. All that *Luke* could do was to fix the beginning of the ministry of John the Baptist and to state that Jesus was approximately thirty years of age when he began his own.

The writings of several early Christian authors from the late second to the mid-fourth centuries contain remarks concerning the year of Jesus' birth. The evidence has been collected and analyzed by Finegan (1964 222–234) and can be tabulated as follows:

Irenaeus: the 41st year of Augustus = Aug. 3–Aug. 2 B.C.[11]
Clement of Alexandria: November 18, 3 B.C. or January 6 2 B.C.
Tertullian: the 41st year of Augustus.
Julius Africanus: Year 5500 from Adam = Olympiad 194, 2 = 3/2 B.C.
Hippolytus of Rome: Year 5502 from Adam = 3/2 B.C.
Origen: the 41st year of Augustus.
Eusebius: the 28th year after the annexation of Egypt and the deaths of Antony and Cleopatra = Aug. 3–Aug. 2 B.C.
Epiphanius: the consulship of 2 B.C.

Although it is surprising to encounter such unanimity in the face of the conflicting chronological data provided by *Matthew* and *Luke*, the consistency in dating Jesus' birth to the year 3/2 B.C. must arise from later Christian writers performing a simple chronological calculation: establishing the year of Jesus' birth simply by going back exactly thirty years from the fifteenth year of the Emperor Tiberius, which results in an annual period beginning in late summer of 3 B.C. and ending in late summer of 2 B.C. In fact, in his long-winded exposition on the subject, in which the huge quantity of words was apparently intended to impress and overawe all readers and critics, Epiphanius (51.22) buttresses his dating of Jesus' birth by listing and counting off thirty years of the Roman consular *fasti* for this very interval. It is therefore apparent that in determining the year of Jesus' birth later Christian writers used the chronological information in *Luke* 3.1–3.2 and 3.23 and ignored the approximate nature of the data. The result, of course, contradicted and invalidated *Matthew's* nativity narrative situated in King Herod's reign, and it was also at variance with Quirinius' census of Judea recorded by *Luke*. Nevertheless, Eusebius, for example, in his *Ecclesiastical History* (I.5–1.8) ignores all such chronological discrepancies. In formulating his own account of Jesus' birth he first dates the event to the year 3/2 B.C., after which he conflates the historically inconsistent Gospel nativity narratives, involving both Quirinius census and King Herod's killing of the babies.

The twenty-first chapter of Book I of the *Stromata* of Clement of Alexandria, written c. 200 A.D., is a lengthy passage that surveys ancient chronology in order to show that Moses and most of the Hebrew prophets preceded the philosophical learning of the Greeks by centuries. Toward the end of the disquisition Clement closes the chronological gap between classical Greece and Hellenistic Egypt, on the one hand, and his own day, on the other, by listing the reigns of the Roman emperors down to the death of Commodus. At the end of this outline, in which he records the reigns of emperors in terms of years, months, and days, he notes that Jesus was born in the 28th year of Augustus. Since Clement is dating this event from the perspective of Egypt that was absorbed into the Roman Empire by Augustus in August of 30 B.C, Clement's 28th year corresponds to 3/2 B.C. Then after making additional remarks regarding the time of Jesus' ministry, Clement (s.147) concludes that the interval of time between the birth of Jesus and the death of Commodus was 194 years, one month, and thirteen days. Since Commodus was murdered on December 31 of 192 A.D., this remark translates into November 18 of 3 B.C. as the actual day of Jesus' birth. Unfortunately, Clement does not enlighten us as to what facts or reasoning underlay such a precise calculation; but he proceeds to record similarly precise calendrical dates (using the Egyptian calendar) in reference to Jesus' life. Thus, according to some, Jesus was born on Pachon 25 (May 20). According to the followers of the gnostic Basilides, Jesus was baptized on Tybi 15 (January 10), but according to others it was four days earlier (January 6). Then after giving different dates for Jesus' death, Clement ends this digression by citing two other groups, one of which thought that Jesus had been born on Pharmuthi 24 (Apr. 19), whereas the other believed it to have been Pharmuthi 25 (Apr. 20). Thus, within the space of a few lines Clement tersely records no less than four different possible birthdays for Jesus.

Two other calendrical dates for Jesus' birth are given by Epiphanius, the bishop of Salamis on Cyprus during the second half of the fourth century. His work, *Panarion* (*Medicine Chest*), argues against the unorthodox beliefs and practices of eighty heretical Christian groups and offers arguments against them. The work was composed c.375. The fifty-first part of this work treats a group of heretics whom Epiphanius terms *Alogoi*, because they did not accept as scripture the *Gospel of John*, which begins with the famous equation between God and *Logos*. In the course of his discussion of this sect Epiphanius (51.22.4) states that Jesus was born "on the eighth day before the ides of January [January 6], thirteen days after the winter solstice and the beginning of the increase of the light and day." Then further on he says (51.29.2) that these *Alogoi* assigned Jesus' birth to the twelfth day before the kalends of June or July (the manuscript reading is uncertain), thus giving a birthday of either May 21 or June 20. Epiphanius, however, in order to reconcile this date with his own, considers the former to be not the date of Jesus' birth, but of his conception; and by assuming a gestation period of only seven months, he is able to make logical sense of the two dates.

The six possible birthdays for Jesus recorded by Clement and Epiphanius could be easily and logically reduced to four. The difference of only one day between Apr. 19 and 20 could be accounted for by assuming that two groups actually agreed upon when Jesus was born, but arrived at two adjacent days in the calendar by having different demarcations for the beginning of a day. For example, if Jesus were believed to have been born during the evening after sunset, one group, who, like the Jews, reckoned the beginning of a day as starting from sunset, would have ascribed the birth to one day; whereas another group, who reckoned the beginning of a day from midnight or from sunrise, would have assigned the birth to the preceding day. Moreover, if May 21, rather than June 20, is accepted as the correct reading of Epiphanius' text concerning the beliefs of the *Alogoi*, Clement's May 20 and Epiphanius' May 21 could also be another instance in which two groups ascribed two different days to the same event, because they had divergent ideas of diurnal time reckoning. If so, the four days (Apr. 19 and 20, and May 20 and 21) could simply represent two days: Apr 19 and May 20 according to Greco-Roman reckoning, or Apr. 20 and May 21 according to Jewish reckoning. If this is the correct explanation for these four different birthdays, they would further provide interesting evidence on how early Christian practices underwent slight temporal shifts as they moved from a Jewish to a gentile cultural environment.

Another interesting aspect of these different birthdays is that the ones in April and May are separated from one another by one month. Even though *Matthew* and *Luke* appear to have had no reliable information concerning the year of Jesus' birth, there might have existed an oral tradition as to what day of the year was Jesus' birthday, because Jesus and his immediate associates must have known this, and they would have known it in terms of the Jewish lunar calendar. But given the temporal fluctuations of such a calendar due to the slippage between the lunar and solar cycles and the need to intercalate, the difference of a month between the two sets of birthdays might have resulted from different methods used in translating a day in the Jewish lunar calendar into other calendars current throughout the eastern Mediterranean, which were quite numerous and varied (lunar, lunisolar, and solar). Moreover, it is noteworthy that these birthdays of April and May differ significantly from the other two (November 18 and January 6), which placed Jesus' birth during the winter. Shepherds being out in the fields with their flocks, as described in *Luke* 2.8, is inconsistent with Jesus' birth in the winter when sheep were kept in their pens, but rather, it fits all other times of the year when weather conditions allowed shepherds and sheep to be outdoors, but the shepherds' vigilance over the animals was especially required in springtime during lambing season, which would be consistent with Jesus' birth in April or May. Thus, *Luke* 2.8 could reflect the actual time of year in which Jesus was born. Alternatively, given the generally unhistorical nature of *Luke's* nativity narrative, dating Jesus' birth to the springtime could have resulted from early Christians, lacking any real knowledge of the circumstances surrounding Jesus' birth, but wanting to fill this void with something substantial, excogitated from *Luke* 2.8 the idea that Jesus had been born during the lambing season in Judea.

As known from later ancient sources, Jesus' birth was widely celebrated on January 6, the day on which, according to some (Clement *Strom.* I.21.146), Jesus was also baptized. This fact served as one of the cornerstones to H. Usener's monumental and epoch-making *Das Weihnachsfest*, first published in 1889 and reissued in a second edition in 1911. Usener's central thesis was that January 6 had preceded December 25 as Jesus' birthday in the Greek eastern half of the Roman Empire; and that it had originated with the gnostic sect of Basilides (early second century), from whom it was adopted by other Christians in the eastern provinces. Usener argued that according to these gnostics, Jesus had been born a mere human, but on the day of his baptism he had become the son of God when the Holy Spirit in the form of a dove descended upon him and entered him. Other Christians, however, modified this doctrine in asserting that Jesus had been born the true son of God on January 6 and was baptized on his thirtieth birthday. In support of this scheme Christians interpreted *Archomenos* (= beginning) in Luke 3.23 as referring to Jesus beginning the thirtieth year of his life, but this rendering is inconsistent with the obvious vagueness of Luke's *hosei* (= about) in specifying Jesus' approximate age and also ignores the more natural meaning of *archomenos* as referring to the beginning of Jesus' ministry. But to return to Usener, his detailed analysis of all the scanty relevant data scattered throughout early Christian literature that has a bearing upon the origins of Christmas underlies all subsequent treatments by modern scholars of early Christianity. According to Usener, by the early fourth century January 6 was being widely celebrated as Jesus' birthday, but from the middle of the century onwards December 25 began to replace January 6 with the new date being instituted by the pope in Rome, whence it gradually spread to become the norm. Despite the previous acceptance of January 6, Usener argued, the new date was adopted in part due to fourth-century Christian theological disputes over the nature of Jesus' divinity. Upholders of the orthodoxy of the first ecumenical council of Nicaea of 325 would have wished to dissociate the day of Jesus' birth from that of his baptism so as to invalidate gnostic and similar theological doctrines that in any way questioned the eternal coexistence of God the Father and God the Son.

Usener's collection of the early Christian data and his analysis thereof are thorough, detailed, and quite compelling. Yet,Roland Bainton (1962) made a very important modification to Usener's overall thesis. He was able to demonstrate that January 6 as Jesus' birthday did not originate with the gnostic followers of Basilides, but it was a notion shared by Montanists, Marcionists, and probably some Orthodox Christians at a relatively early date. Instead of concluding that the idea had begun with the Basilidians and spread to other Christians, Bainton argued that given the different groups' antagonism toward one another, it was more likely that their shared tradition of January 6 went back to a time before they had diverged into different religious sects, thus taking January 6 as Jesus' birthday back to the early second century. Bainton further made the astonishing observation that November 18 of 3 B.C., so exactly recorded by Clement, actually corresponds to January 6 of 2 B.C. if one employs the Nabonazzar calendar that consists of a 365-day year without the addition of the leap-year day every four years. For the 194 years between Jesus' birth and the death of Commodus the difference between the Julian and Nabonazzar calendars is 48 or 49 days, thus advancing November 18 of the Julian calendar to January 6. Consequently, November 18 as Jesus' birthday turns out to simply be January 6 by a different mode of reckoning. The important conclusion to be reached from all this is that when Clement was writing c.200, several different Christian sects were in agreement in assigning Jesus' birth to January 6, and that this tradition probably went back as far as the early second century.

What might have been the reason for such a date? Epiphanius (51.22.9) records a pagan celebration observed at Alexandria during the night of January 5/6. It took place at the *Koreion*, a shrine of Kore, the maiden daughter of Demeter. Worshippers held a vigil, singing hymns accompanied by flutes. It ended at dawn when there was carried into their midst from an underground chamber a divine image, naked and lying on a litter. It symbolized Kore giving birth to Aion, the god of eternity. Epiphanius regarded these mystic rites as a pagan adulteration of the true Christian ones, designed to lure the unsuspecting into falsehood. This was the common attitude adopted by Christian theologians when confronted with similarities between Christian and pagan practices. Since mystic rites of this sort had long been a tradition in Egypt, the coincidence of Aion and Jesus being born on the same day is certainly to be explained as resulting from Christian imitation of the pagan rites rather than vice versa.[12] These mystic rites in Alexandria can also account for the origin of the term Epiphany used by Christians as the name for January 6. The word simply means "manifestation" and was a common religious concept of Hellenistic religious thought, involving a divinity manifesting himself in a visible form to human beings. The term therefore aptly describes the ceremony recorded by Epiphanius and is likely to have been adopted by the Christians of Alexandria along with their version of the mystic rites themselves.

In conclusion, the surviving ancient evidence indicates that around 200 A.D. Jesus' birth was clearly not associated in any way with the winter solstice. Rather, the testimony of Clement of Alexandria along with other data shows that many Christians of the Greek East regarded January 6 as the day of Jesus' birth and may have already been commemorating it under the name Epiphany.

EQUINOXES AND SOLSTICES IN THE RELIGION OF THE REPUBLIC

In his description of the Roman agricultural year Varro is well aware of the equinoxes and solstices, but unlike our modern practice of dating the arrival of the four seasons to these turning points, Varro (*De Re Rust.* I.28) places the equinoxes and solstices at the midpoints of the seasons. Of these four major turning points in the annual solar cycle, the summer solstice alone, occurring on June 24 in the Julian calendar in ancient Roman times, was distinguished by an important cultic celebration.[13] The day was sacred to Fors Fortuna, and the rites of the day centered around two temples to the divinity ascribed to King Servius Tullius, located downstream on the Tiber from Rome and on the opposite side of the river, and situated at the first and sixth milestones along the Via Portuensis (*Fasti Amiternini* = Degrassi 1963 187). According to Ovid (*Fasti* VI.771–784, cf. Cic. *De Fin.* V.70) the festival was a popular celebration characterized by people thronging the river bank, sailing in boats up and down the river, and indulging in much drinking. The day was later consecrated by the Catholic Church as the birthday of John the Baptist, which Sir James George Frazer (1929 IV. 332–335) regarded as the Church's attempt to Christianize a popular and widespread European pagan celebration of Midsummer Day, on which people bathed, because waters on that day were believed to possess special curative powers:

> . . . and that the same day (Midsummer Day or Midsummer Eve) appears to have been a very ancient festival of water in Europe, especially in southern Europe, which may have suggested to the Church the propriety of placing Midsummer Day under the patronage of St. John the Baptist, thereby throwing a Christian cloak over an old heathen celebration. The European Midsummer festival, like the Roman festival described by Ovid, has been essentially a popular holiday. Water is then supposed to acquire certain marvelous medicinal properties, and people seek to take advantage of them by bathing in the sea, rivers, or springs, or rolling in the dew. To roll in the Midsummer dew is esteemed especially a cure for diseases of the skin. Hence in many parts of Europe, from Sweden in the north to Sicily in the south, and from Ireland and Spain in the west to Esthonia in the east it used to be customary for men, women, and children to bathe in crowds in rivers, the sea, or springs on Midsummer

> Eve or Midsummer Day, hoping thus to fortify themselves for the next twelve months. . . . It might perhaps be thought that this widespread custom of bathing in water or dew on Midsummer Eve or Midsummer Day is of purely Christian origin and has been adopted as an appropriate way of celebrating the day dedicated to St. John the Baptist, who had enjoined such dips on all his disciples. But two considerations seem fatal to this view. In the first place, the custom was denounced and forbidden as a heathen practice by St. Augustine. In the second place the custom is observed to this day by Mohammedan peoples of North Africa, particularly of Morocco, who have no respect for St. John the Baptist and no desire to follow his precepts. These Moslems of Africa, like the Christians of Europe, believe all water to be endowed with such marvelous virtue on Midsummer Day that it not only heals sickness but prevents it for the rest of the year; hence men, women, and children bathe in the sea, in rivers, or in their houses at that time for the sake of their health. Thus we seem justified in concluding that the custom of bathing in water at this season of the year as a remedy for or preventative of disease is part of an old heathen celebration of Midsummer, which was once, and to some extent still is, common to the Christian and Mohammedan peoples on both sides of the Mediterranean. It is possible that the aquatic festival at Rome on Midsummer Day is to be classed among these Midsummer rites of water, and that the revellers on that day bathed or washed in the Tiber as well as floated on its surface.

As best we can tell, the other three annual turning points of the solar cycle did not receive any special cultic attention in the official calendar of the Roman Republic or early Empire. Nevertheless, the two equinoxes and the summer solstice might have been associated with Minerva. The month of September in the Roman calendar was totally barren of festivals except for the anniversary of the day of dedication of the Capitoline temple on the ides. Since the temple in question was that of Jupiter Optimus Maximus, the god of the sky and of daylight, we might suppose that there was a connection between the deity and the autumnal equinox; and that Roman priests, rather than assigning the day of dedication to the equinox, preferred instead to assign it to the ides of the month, because all ides were sacred to Jupiter. In his well-known description of the early Republican custom of the *clavus annalis*, performed every year on September 13, Livy (VII.3.5–6) states that the nail was driven into the wall of the cella of Jupiter Optimus Maximus that faced the cella of Minerva, because these nails were used to mark the passage of the years in early times, and "number" had been the invention of Minerva. Thus, Livy's account associates this annual ceremony, conducted shortly before the arrival of the autumnal equinox, with Minerva. Likewise, the other two important festivals to Minerva in the Roman calendar were the Quinquatrus of March 19 and the Quinquatrus Minusculae of June 13, both coming shortly before the vernal equinox and summer solstice respectively.

Although mid-December contained several festivals (Consualia on the 15th, Saturnalia on the 17th, and the Opalia on the 19th), they were all connected with the ending of the Roman farmer's winter planting, and there is nothing in them to suggest an association with the winter solstice. Unfortunately, the Divalia and Larentalia (or Larentinalia) of December 21 and 23 had become so obscure by Varro's day that the Romans had no clear understanding as to their original nature and meaning. Thus, even though there might have been in Republican times some vague connection between Minerva and three of the four annual solar turning points, as suggested in the preceding lines, they must have been quite tenuous at best, and we can conclude rather confidently that apart from the popular celebration of Fors Fortuna on the summer solstice, Roman religion of the Republic and early principate did not have strong ties to the equinoxes and solstices.

A somewhat similar picture emerges in reference to the Roman worship of the Sun, because Sol as an independent divinity, not equated with Greek Apollo, does not seem to have been very prominent in early Roman cult. According to Varro (*L. L.* V.68 and 74) Sol was Sabine in origin and among other deities was introduced into Rome by T. Tatius (cf. Dion Hal. II.50.3 and Aug. *Civ. Dei* IV.23). Festus (22L s.v. *Aureliam*) associates the worship of Sol with the Aurelian family of Rome. Since the Etruscan name of the sun god was Usil, modern scholars have plausibly conjectured that the early unrhoticized form of Aurelius (= Auselius) derives from this Etruscan divine name and thus explains Festus' connection between the Aurelii and the worship of Sol (see Radke 1979 289).

Rome during Republican times seems to have had only two precincts dedicated to the Sun, but both were apparently of great antiquity. One shrine was situated in the valley of the Circus Maximus and was later incorporated into the structure of the seating, as shown on Imperial coins.[14] It was probably situated near the finish line, where Sol, charioteer of the sky, could observe and preside over the end of the races. Tacitus (*Ann.* XV.74) terms it a *vetus aedes* and mentions it as the place where the Pisonian conspiracy against Nero was formed; and when it was uncovered and successfully suppressed, sacrifices were performed to all-seeing Sol as the revealer of the plot. The day of dedication for this shrine from the early principate onwards seems to have been August 28 (Degrassi 1963 135 and 253).

The other precinct to the Sun was located on the Quirinal. In discussing the peculiar features of archaic Latin Quintilian (I.7.12) mentions that at the Pulvinar of Sol next to the temple of Quirinus there was to be found an inscription containing the form *Vesperug* in place of *Vesperugo* (= the evening star). It is certainly unfortunate that we do not know much more about this archaic inscription, because it clearly suggests that it was somehow astronomical in nature. It therefore should be viewed in connection with Pliny *NH* VII.213, which cites Fabius Vestalis for the claim that the first sundial set up in Rome was by L. Papirius Cursor in 293 B.C. when he dedicated the temple of Quirinus (see Livy X.46.7). Pliny, however, dismisses Fabius Vestalis' testimony and instead sides with Varro in ascribing Rome's first sundial to M'. Valerius Messala, who as consul in 263 B.C. set up next to the Rostra a sundial taken as a war trophy from Catana in Sicily. Since T. Tatius and his Sabine followers were supposed to have settled on the Quirinal after their war with Romulus, Quintilian's *Pulvinar Solis* and Fabius Vestalis' sundial were probably part of an open-air precinct (*sacellum*) containing an archaic altar dedicated to the Sun that was later attributed to T. Tatius.

The epigraphic calendars (Degrassi 1963 149, 181, and 191) record August 9 as the day sacred to Sol Indiges on the Quirinal. It is noteworthy that this day was neither an equinox nor a solstice, thus corroborating the conclusion reached above that in matters of religion the early Romans were relatively indifferent to the four turning points of the annual solar cycle. Rather, August 9 fits the larger pattern of early Roman festivals in July and August in displaying the peasant farmer's concern to protect his land, crops, and forests from the relentless summer sun and their possible destruction by fire. The Lucaria of July 19 and 21, dedicated to sacred groves, must have represented the farmer's need to clear away dry wood and all other such material to prevent the outbreak of forest fires. The Neptunalia came two days later on July 23 and was dedicated to the god of fresh-water streams, springs, and pools at a time when drought would have been a real threat. Then following August 9 sacred to Sol Indiges, the Volcanalia of August 23 closed out this series of interrelated festivals designed to protect the Romans from the danger of fire resulting from the dry conditions of summer.[15]

OBELISKS AND EGYPTIANIZED SOLAR WORSHIP

As in so many other things, the major transition in solar worship among the Romans was brought about by Augustus. In 10 B.C. two huge Egyptian obelisks were transported to Rome, reerected, and officially dedicated to Sol.[16] Egypt had always been regarded as the cradle of civilization by the Greeks and Romans, who therefore viewed its religious traditions as most ancient and, hence, most venerable and efficacious. For the past 3000 years solar worship had been integral to Egyptian culture, and during the brief reign of Akhenaten in the fourteenth century B.C. it had even formed the basis of the earliest monotheism known to history. From the beginning of their civilization the Egyptians had employed a solar calendar consisting of 365 days, later revised under Ptolemy III Euergetes to include a leap-year day every fourth year (see *OGIS* 56). It was this solar calendar, borrowed from Egypt, which Caesar had instituted in 46/45 B.C. as Rome's new Julian calendar.[17]

Since the time of the Middle Kingdom (c.2000 B.C.) Egyptian pharaohs had been erecting pairs of obelisks to flank the entrances to temples. They consisted of a single, solid shaft of stone quarried from granite. Their shape, having a square cross-section that tapered slightly from bottom to top and terminating in a small pyramid, represented a sunbeam of the sun-god Re, to whom all obelisks were sacred. In addition, the small pyramid at their top was called the benben and symbolized the mound of creation, which according to Egyptian mythology had emerged from the primordial sea, and upon which the god Ptah had brought about the creation of all living things. The benben was usually sheathed in gold or bronze, so that as the sun arose each morning, it was the first thing to catch the light of the new day. Successfully transporting two of these enormous objects from Egypt to Rome by sea aboard ships specially constructed for the job and then reerecting the obelisks in Rome were demonstrations of Roman power, superior engineering, Roman mastery of Egypt, and Augustus' ability to appropriate the venerable cultural legacy of Egypt and to reapply it to Rome's own uses and needs. In fact, the ships used to transport these enormous monuments were such marvels of Roman engineering that Augustus kept one on permanent display at Puteoli; and Claudius used a similar barge built under Caligula for transporting another obelisk as part of the break-water structure for the harbor at Ostia (Pliny XXXVI.70).

One of the two Augustan obelisks, measuring about 80 feet high, weighing about 400 tons, and dating to the early thirteenth century B.C. during the reign of Ramesses II (now in the Piazza del Popolo), was erected at the midpoint of the *spina* of the Circus Maximus (Humphrey 1986 271). Its placement was also designed to add grandeur and distinction to the old shrine of Sol that had long been incorporated into the fabric of the Circus Maximus. In order to signify Rome's formal appropriation of this Egyptian monument of solar worship, Augustus had the following Latin inscription cut upon two opposite faces of the obelisk's base, which many spectators in the Circus Maximus could have seen and read: "After Egypt had been reduced to the power of the Roman people, the Emperor Caesar Augustus, son of the deified [Julius], *pontifex maximus*, hailed *imperator* twelve times, eleven times consul, in the fourteenth year of his tribunician power, gave [this] as an offering to Sol."[18] The obelisk's day of dedication is likely to have been December 11, because according to the *Fasti Ostienses* and the *Fasti Amiternini* (Degrassi 1963 106 and 199) this day was sacred to *Indiges*. Johannes Lydus, writing in the sixth century of our era (*De Mensibus* IV.155), states that on December 11 the Romans performed rites to *Genarches Helios* (= Sol Indiges). Of the three instances in which Sol is recorded in the epigraphic calendars, this one is the closest to a solstice, just fourteen days before December 25.

The Christian polemicist Tertullian, writing in his *De Spectaculis* (8–9) c.200 A.D., provides us with interesting information concerning Sol and the chariot races conducted in the Circus Maximus. In advising his fellow Christians to avoid these celebrations because of their idolatry, he indicates that the temple of Sol was situated at the center of the race course with Sol's cult statue on top of the structure. He further notes that the four-horse chariots were all sacred to the Sun, and the two-horse chariots were dedicated to the Moon. Moreover, in the penultimate chapter of the treatise (29), which forms the work's peroration, Tertullian likens Christian devotion to the pleasures of the *ludi circenses*, which he terms *cursus saeculi, tempora labentia, spatia peracta*, and *metas consummationis*. His language suggests that the laps of the chariots around the Circus Maximus were regarded as symbolizing the passage of time as marked by the cycles accomplished by the sun and moon. Indeed, since all chariot races consisted of seven laps around the track, the Romans of Imperial times must have associated the races with the seven ancient planets and the seven days of the planetary week discussed in the next section of this chapter. Thus, in Tertullian's day the celebrations in the Circus Maximus were imbued with what we might term a solar ideology, in which Augustus' obelisk must have played a prominent part.

The second Augustan obelisk, measuring about 72 feet in height, had been hewn from red granite and erected at Heliopolis during the early sixth century under Psammetichus II. When reerected in Rome, its base was inscribed with the same Latin text just quoted from the other Augustan obelisk, thus indicating that it too was a formal dedication to Sol. The purpose to which Augustus put this monument was, however, even more grandiose than that of the other obelisk, truly befitting not only an Egyptian pharaoh, but also the founder of the autocratic principate. The obelisk was erected in the Campus Martius and served as the gnomon of an enormous sundial, which Pliny (XXXVI.72) describes as follows: "To this one, which is in the Campus, the deified Augustus added a marvelous use for catching the shadows of the sun and thus the lengths of the days and nights with a stone pavement for the obelisk's length, to which its shadow became equal at noon on the day of winter's beginning and gradually decreased and again increased day by day along the lines inlaid in bronze."[19] This brief description of Augustus' colossal sundial has been strikingly confirmed and further amplified by Edmund Buchner. During archaeological soundings in the Campus Martius in 1979 and 1980 Buchner discovered parts of the pavement for the sundial, as well as a few of the large bronze letters used to label the signs of the zodiac in Greek; but even more astonishing was Buchner's earlier work. He had supposed that there existed a complex relationship between the sundial and the other two nearby Augustan monuments, the Mausoleum Augusti and the Altar of Augustan Peace. Buchner's careful consideration of their locations and his calculations have led him to conclude that at the autumnal equinox the obelisk's shadow mounted the steps and entered the inner part of the *Ara Pacis*, and on the winter solstice the shadow fell upon the Mausoleum (Buchner 1982 36). The former phenomenon, Buchner concluded, was designed to acknowledge Augustus' birthday on September 23, whereas the latter celebrated his conception. Noteworthy in this regard is a decree from the Roman province of Asia and dating to 9 B.C., the very next year after the formal dedications of the two Augustan obelisks to Sol.[20] The inscription records the decision of the Greek cities of the province, acting in response to the suggestion of the provincial governor Paulus Fabius Maximus, to alter their calendar, so that Augustus' birthday would henceforth become the first day of the year.

Augustus' adaptation of Egyptian solar worship was typical of the heterogeneous religious culture of Rome that had been gradually developing since the beginning of the second century B.C. As Rome emerged as the ruling power of the Mediterranean, the city grew by leaps and bounds, and its population was increasingly composed of foreign residents and slaves and freedmen of foreign birth, who brought into Rome their native culture and religious beliefs and practices. This cultural diversity, however, often incurred the distrust and hostility of the Roman upper class. The senate's well-known crackdown on Bacchic worship in 186 B.C. (see Livy XXXIX.8–19 with *ILLRP* 511 = Gordon 1983 #8) is our earliest and best attested illustration of this phenomenon. By the first century B.C. the worship of Isis and Serapis, the chief Hellenistic divinities of Egypt, had become quite popular in Rome, so much so that at the time of Sulla's dictatorship they were sufficiently numerous to have a *collegium* of *pastophori* (Appuleius *Metam.* XI.30). Catullus (10.26), writing during the 50's B.C., testifies to the existence in the city of a temple to Serapis, which was a popular hangout for prostitutes and other unsavory folk. As in the case of the Bacchanalia, many members of the Roman elite took a dim view of the popularity of Isis and Serapis among the masses; and during the years 58–48 B.C. the senate and magistrates of Rome repeatedly attempted to suppress the cult by destroying its altars and temples.[21] In 43 B.C., however, when the triumvirate of Antony, Lepidus, and Octavian came to power, they decreed the construction of a temple to Isis and Serapis, apparently in an attempt to garner much needed popular support of their ruthless regime (Dio XLVII.16.1). Yet, despite Augustus' willingness to appropriate Egyptian obelisks for his own aggrandisement, his principate's official attitude toward the gods of Egypt was somewhat tempered. In 28 B.C., three years after the battle of Actium had pitted Italy led by Octavian against Egypt headed by Antony and Cleopatra, Octavian decided that the Egyptian gods were not to be worshipped inside the *pomerium*, and seven years later in 21 Agrippa extended this area to include a distance of one mile from the *pomerium* (Dio LIII.2.4 and LIV.6.6). Tiberius continued this traditional policy of Augustus toward the gods of Egypt; and in 19 A.D., as the result of a scandal involving the unwitting seduction of a Roman matron in a temple of Isis, the senate and Tiberius took extremely harsh measures against the cult in trying to extirpate it from the city.[22]

A new receptive policy was probably introduced by Gaius Caligula. Although the ancient evidence is by no means certain, modern scholars have generally inferred from it that Caligula is our most likely candidate for building a splendid temple to Isis and Serapis in the Campus Martius (see Wissowa 1912 353–354 and Richardson 1992 211 with Colin 1954). On Caligula's orders another obelisk was transported to Rome (Gordon 1983 #35). Augustus had intended to bring it to Rome, but Cornelius Gallus, the first Roman equestrian prefect of Egypt, had moved it from Heliopolis to Alexandria, where he had it erected in the Forum. When it arrived in Rome, like one of the Augustan obelisks, it was set up on the *spina* of a race course, the so-called

Circus Gaii et Neronis in the private gardens of the Caesars in the Vatican. There the obelisk stood for over 1500 years just to the south of the southern transept of Constantine's Basilica of St. Peter until in 1586 it was moved to its present position in the center of the Piazza S. Pietro in front of the modern-day Basilica of St. Peter.

Other Egyptian obelisks were brought to Rome during the reign of Domitian. Indeed, the Flavian dynasty (69–96 A.D.) represented Rome's full acceptance of Egypt's gods, in part because Egypt had played a very crucial role in elevating Vespasian to the imperial throne. The Roman troops stationed at Alexandria were the first to swear allegiance to Vespasian on July 1 of 69 A.D., thereafter celebrated by him as his *dies imperii* (Tac. *Hist.* II.79). Shortly thereafter, while visiting Alexandria, Vespasian was thought to have brought about two miraculous cures after the afflicted persons had been urged by Serapis in a dream to seek out the new emperor. Vespasian then worshipped Serapis in his famous temple in Alexandria (Tac. *Hist.* IV.81–82). The emperor's special regard for the Egyptian god is indicated by the fact that the night before entering Rome to celebrate his triumph over the Jews, he and his older son Titus spent the night in the temple of Isis and Serapis in the Campus Martius (Josephus *Bell. Iud.* VII.5.4). When Vespasian constructed his Forum (Templum Pacis) out of the booty from the Jewish War, it was adorned with many works of art, including a stone representation of the Nile surrounded by sixteen children, whose number derived from the number of cubits measured for a high inundation of the river (Pliny XXXVI.58). The Emperor Domitian, Vespasian's younger son, seems to have been equally pious toward the gods of Egypt. When the temple of Isis and Serapis burned down in 80 A.D. during the brief reign of Titus (Dio LXVI.24.2), Domitian not only rebuilt the shrine (Eutrop. VII.23.5), but he adorned it with obelisks brought from Egypt (Richardson 1992 274–275). The new sacred precinct was a large rectangular area enclosing two temples, one of Isis to the north and another of Serapis to the south, joined in the center by a courtyard, in which probably stood the obelisk that now stands in the center of Bernini's Fountain of the Four Rivers in Piazza Navona. The obelisk bears a Domitianic inscription written in Egyptian hieroglyphics, informing us that the stone shaft was dedicated to Harakhte, the Egyptian god of the rising sun, who is termed the emperor's father. The inscription also alludes to the death and deification of Titus (Iversen 1968 78–80). Thus, by 100 A.D. Rome had become the welcome home of Isis and Serapis and had incorporated into its urban landscape and architecture Egyptian obelisks that were well-known cultic objects of solar worship.

ROMAN ADOPTION OF THE SEVEN-DAY WEEK

As already mentioned in Chapter 1, down to the end of the Republic the Romans organized their lives around an eight-day week (*nundinae*), by which they worked for seven days and used the eighth day for attending markets and public meetings. This eight-day cycle was enshrined in the official Roman calendar by a series of nundinal letters A–H. By the time of Augustus, however, we begin to have evidence that a seven-day week was beginning to gain currency alongside the eight-day one. Like the Julian Calendar, this seven-day cycle was of eastern origin, but unlike Caesar's calendrical reform, no official act was ever taken by the Roman state to replace the older with the newer weekly cycle. Through a process that is unfortunately inadequately documented by our surviving evidence, the traditional nundinal week slowly gave way to the seven-day one.[23]

The *Fasti Sabini* (Degrassi 1963 51–54), dating sometime after 19 B.C., is preserved only for a little more than half the months of September (7–25) and October (10–27), but what does survive offers important information about the seven-day week. Instead of having a single column of nundinal letters A–H, there is a series of double letters: the first one of the pair forming a cycle of A–G followed by a second letter from the traditional nundinal sequence, thus marking off both seven-day and eight-day weekly cycles. The exact same pattern of double letters, one for a seven-day week immediately followed by one for the eight-day week, occurs in two other fragmentary calendars: the *Fasti Foronovani* and the *Fasti Nolani* (Degrassi 1963 156 and 229–231), both dating to the early principate. Since in all three of these fragmentary calendars the letters for the seven-day week precede the traditional nundinal letters, we might tentatively conclude that the former were regarded as the norm, and the latter were included for the sake of tradition, or to show how a newly accepted system matched up to the older one that it was replacing, and which was passing out of existence. If so, the three *fasti* would constitute solid evidence for the growing acceptance of the seven-day week during the early years of the principate.

Brief allusions in both Ovid and Horace suggest that many Romans during Augustan times were well aware of the Jewish custom of observing the Sabbath every seventh day.[24] We might therefore suppose that the seven-day week adopted by the Romans was of Jewish origin, but additional considerations of our surviving evidence present a much more complex picture. The Jews did not assign names to the seven days of their week, but they simply numbered them, as we encounter them in the opening verses of *Genesis* in God's creation of the universe (cf. *Exodus* 20.8–11). On the other hand, the days of the Roman seven-day week, which slowly won acceptance during Imperial times, took their names from the seven ancient planets in the following order: Saturn, Sol, Luna, Mars, Mercury, Jupiter, and Venus, thus forming a so-called planetary week. Consequently, Colson (1926 43–56) concluded that the Roman planetary week grew out of Hellenistic astrology that became increasingly pervasive from c.200 B.C. onwards. Indeed, Webster (1916 223ff, cf. Bickerman 1980 59) cites evidence for the existence of a seven-day cycle among the Sumerians and Babylonians of the third and second millennia B.C. It is therefore likely that Mesopotamia was the original homeland of the seven-day week; and as in so many other things, the Hebrews, who did not arrive on the historical scene until c.1000 B.C., drew upon this ancient Near Eastern tradition and adapted it in their own way. Then, following Alexander's conquest of the Persian Empire, the Near Eastern seven-day week became part of the Hellenistic culture that gradually spread westward. Thus, the Jewish and planetary weeks were simply two variations upon the same larger tradition.

Tibullus I.3.18 contains our earliest extant reference in Roman literature to a day of the planetary week: "Saturnive sacram me tenuisse diem = or that the sacred day of Saturn held me." Amid a list of omens that did not prevent Tibullus from leaving his beloved Delia to join Messalla as a member of his entourage to the provinces, the poet includes the taboo nature of Saturday, but rather than referring to the Jewish Sabbath itself as Ovid and Horace do, the day is here instead termed *Saturni dies*, the equivalent of the Jewish Sabbath in the Roman planetary week. As a result of this equation, the monotheistic god of the Jews was sometimes identified by Imperial authors as Saturn (see Tac. *Hist.* V.2–4 and Dio XXXVII.16–17). Three inscriptions from Pompeii (*CIL* IV. 4182, 6779, and 8820) indicate that by the time of the city's destruction in 79 A.D. the planetary week had gained wide acceptance and had become an integral part of the popular culture. The first of these items is a short inscription listing the market days for Pompeii and Cumae. It bears the consular date 60 A.D. and along with a traditional Roman calendrical date further supplies "*dies Solis, Luna[e]* = Sunday and Monday." The second inscription is a simple list of the seven planetary days of the week, inscribed in a vertical column and giving the names of the planets in the genitive case without the pleonastic *dies*: Saturni | "Solis | Lunae | Martis |—| Iovis | V[e]n[e]ris."[25] The third inscription is a *graffito* scratched by a stylus onto the wall of a bedroom. It reads: "X K(alendas) Febra. Ursa peperit diem Iovis = on Thursday January 23 Ursa gave birth." The commonplace nature of this last item, in which a planetary day of the week is given alongside the traditional Roman calendrical date for an important event in the life of a family, clearly shows that the planetary week had permeated the social fabric of Pompeii. The latter conclusion is borne out by a remark of Josephus (*In Apionem* II.39). Writing toward the end of the first century, the Jewish author observed that the seven-day week had become universally accepted; and believing it to be of Jewish origin, he regarded this phenomenon as everyone's tacit approval of the Jews and of their institutions.

According to *Matthew* 28.1 Jesus arose from the dead at dawn on the first day of the Jewish week, following the seventh day, the Sabbath. As a result, early Christians observed this day, not the seventh, as their holy day of the week; and by the end of the first century (and possibly from the very outset) they were calling it "the Lord's Day," as we see in *Revelation* 1.10. Similarly, Ignatius, the bishop of Antioch, in a letter written during the early second century (*Ad Magnesias* 9), comments that Jews who have converted to Christianity no longer observe the Jewish Sabbath on the seventh day, but instead, worship in accordance with the Lord's Day. In his defense of Christianity addressed to non-believers, written around the middle of the second century, Justin Martyr (*Apologia* I.67) explains that his fellow Christians gather on Sunday to read the Gospels and to partake of the eucharist, because it was on this day, the first day of the Jewish week, that God began creating the universe, and Jesus arose from the dead, which was the source of their salvation. It is therefore clear that just as the Jewish Sabbath was equated with Saturn's Day in the planetary week, so the Lord's Day of the Christians must have been identified early on as the *dies Solis* of the planetary week. In fact, this is confirmed by Tertullian (*Apol.* 16.9–11 and *Ad Nationes* I.13), because Christians worshipped their savior on Sunday and, when praying, faced eastward toward the rising sun. Besides, equation between the sun and the Christian god was in accord with Christian belief, because Jesus, *inter alia*, was viewed by Christians as the light that had shown into the darkness (*John* 1.4–5).

SOL INVICTUS IN THE ROMAN EMPIRE

The next major phase in the history of solar worship in Roman religion concerns not Sol, but Sol Invictus. Like the Hellenistic period that had gone before, the Roman principate produced a religious culture of bewildering diversity, which by the second century was characterized by syncretism.[26] The *locus classicus* for this phenomenon is, of course, Isis' famous address to Lucius at the beginning of Book XI of Apuleius' *Metamorphosis*, in which the goddess explained that all the major female divinities worshipped by the different nationalities of the Roman Empire were simply local manifestations of the same all-powerful universal goddess, whose true name, known of old by the Egyptians alone, was Isis. As the second and third centuries unfolded, the same syncretism resulted in what we might term solar henotheism, in which the major solar divinities worshipped by different peoples of the Empire were identified with one another and were amalgamated into one divinity, who was elevated to the supreme position in the polytheistic pantheon.[27] Indeed, solar worship in later Imperial times played an important role in the spread and acceptance of Christianity and was absolutely crucial in causing December 25 to become Jesus' birthday.

As demonstrated in the first two sections of this chapter, Sol was a divinity dating back to Rome's earliest times, but he did not figure prominently in the religion of the Republic. Then under Augustus a much more grandiose form of solar worship from Egypt was grafted onto him. Until recently, modern scholars of Roman religion have generally agreed that Sol Invictus, who played such an important role in the religious history of later Roman Imperial times, was of Syrian origin, thus representing the triumph of East over West; but experts in ancient Syrian culture have succeeded in casting doubt upon this long-held doctrine.[28] Accordingly, by drawing upon this recent scholarly research, this treatment of the non-Christian origin of Christmas will, at least in this regard, offer a historical reconstruction of events significantly different from earlier modern accounts.

From the first century of the principate onwards Sol was represented rather uniformly in one of three ways on pottery, reliefs, and coins: (1) male bust wearing a radiate crown or nimbus; (2) standing male, usually wearing radiate crown or nimbus, with one hand upraised in benediction or salutation and the other hand holding a whip used to drive his chariot; and (3) male figure, usually wearing a radiate crown or nimbus and riding in a four-horse chariot.[29] He was often paired with Luna to symbolize night and day. Hijmans (1996 125ff) has argued cogently that Sol's iconography owed nothing to Syria, but instead, was the product of Greco-Roman tradition: for these various motifs are attested in Apulian vase painting of the late classical and Hellenistic periods.

Perhaps the most exceptional and unusual representation of Sol dates to the reign of Vespasian. Following the great fire of 64 A.D., which destroyed a huge portion of Rome, Nero used the opportunity to erect near the Roman Forum a splendid palatial complex known as the Golden House (*Domus Aurea*), in whose courtyard he had erected a gigantic statue of himself, standing over 100 feet tall (Suet. *Nero* 31). Following Nero's death and his own rise to imperial power, Vespasian demolished the Golden House and converted the statue into one of Sol by placing on its head a radiate crown with seven rays twenty-three feet in length (Suet. Vesp. 18 and Pliny NH XXXIV.45). The statue was known as the Colossus in imitation of the famous Colossus of Rhodes erected during the early third century B.C. by the Rhodians to commemorate their victory over Demetrius Poliorcetes. This latter statue also measured slightly more than one-hundred feet high and was dedicated to the sun god Helios, the tutelary divinity of Rhodes (Pliny *NH* XXXIV.41 and Strabo XIV.2.5). As the result of Hadrian moving the Colossus so as to stand near the Flavian Amphitheater (*Hist. Aug. Hadrian* 19.12–13), the latter began to be called the Colosseum.

Following sporadic appearances on coins of the Republic, Sol was not displayed on Imperial coinage until the reign of Vespasian, then of Trajan and Hadrian, and then from Commodus continuously up through the first half of Constantine's reign.[30] It seems that Sol acquired his epithet Invictus during the second half of the second century, at which time he began to emerge as the well-known mighty solar divinity of the Roman Empire. *CIL* VI. 717 (= *ILS* 4217), dating to the year 158 A.D., is our earliest extant text that associates Invictus with Sol. The inscription was cut on a statue base to the god and records that it was a dedication made by two individuals, one of them being the very first master (*magister primi anni*) of a newly founded club (*sodalicium*), whose patron deity was Sol Invictus. Toward the end of his reign, as his megalomania increased, Commodus became the first Roman emperor to adopt the title Invictus, and along with his other imperial names and titles he had it applied to one of the months of the year (Dio LXXIII.15.3). In addition to styling himself *Hercules Romanus* (*Hist. Aug. Commodus* 8.5),he had the Neronian Colossus fitted out with a club to resemble Hercules and with a new head and face resembling himself (Dio LXXIII.22.3). Since at least mid-Republican times the Romans had called Hercules both Victor and Invictus, and the latter epithet of the god was now so closely associated with Sol that Commodus had the colossal statue of Sol Invictus changed to Hercules Romanus Invictus. Caracalla followed in Commodus' footsteps in adopting the title Invictus, and it henceforth became standard usage by all later emperors (Weinstock 1957 243), thereby indirectly associating the emperor with the all-seeing sun. Thus, by the beginning of the third century Roman Sol had become Sol Invictus, a mighty divinity who looked down upon and watched over the vast extent of the Roman Empire.

Mithraism is likely to have played an important role in enhancing the stature of Sol from the second century onwards. Mithras was originally an Indo-Iranian solar divinity, whose cult in eastern Anatolia was Hellenized and shaped into one of the most popular mystery religions of Imperial times. We first learn of this cult at the time of Pompey's war against the pirates in 66 B.C. Plutarch (*Pompey* 24) states that many of the pirates, who came from Cilicia, worshipped Mithras and were the first ones to spread his cult into other areas. Nevertheless, our earliest epigraphic and archaeological evidence for Mithraism as a mystery religion dates to the second century of our era. By that time shrines of Mithras (Mithraea), in which his worshippers gathered to celebrate his mysteries, were being built in Rome, Ostia, along the Rhine and Danube frontier, and in other parts of the empire. Despite the survival of numerous inscriptions relating to the cult and the discovery of many Mithraea decorated with artistic representations pertaining to the mysteries, many questions surrounding the cult still remain unanswered. Indeed, during the past thirty years there has arisen a highly specialized subfield of Mithraic studies within the larger area of the religious history of the Roman Empire; and a small cadre of scholars has been devoting their entire professional careers in studying intensively Mithraism and all its artifacts in an attempt to unlock its secrets.[31]

All Mithraea had as their central focus the tauroctony, Mithras' slaying of the bull, represented in the form of a wall painting, a wall relief, or a sculpture group. It shows the god kneeling on the back of a bull and stabbing it in its chest with a knife. Other figures are also usually included: a dog near the bull's wound, from which blood flows; a scorpion and snake underneath the bull; a raven above Mithras; and two human figures, named Cautes and Cautopates, flanking each side of the scene, standing with their legs crossed and holding a torch, Cautes' held upright and Cautopates' held upside down. The bull's tail is sometimes shown sprouting stalks of grain, and the upper corners of the scene often contain images of the Sun and Moon. Although the tauroctony is still far from being entirely understood, Mithraic experts since the mid 1970's have come to realize that the scene is a kind of star map, illustrating various stars and constellations, such as Spica (the bull's tail), Canis

Minor (the dog), Hydra (the snake), and Scorpio (see Beck 1984 2081–2083 and Ulansey 1989 15–24). Recent scholarship has demonstrated that Mithraism was permeated with astrological ideas. The communities of worshippers were organized into seven grades (Jerome *Ep.* 107.2), each of which was associated with one of the seven ancient planets. Porphyry, a Neoplatonic philosopher of the third century, says that Mithraism taught its initiates about how souls descended and arose again (*De Antro Nympharum* 6). The idea certainly must be that each person's soul, originating as a divine spark in the outermost sphere of the universe, descends through the seven heavenly spheres in order to come to earth to enter its mortal body; and then after death, the soul ascends again. In any case, the distribution of Mithraea along Rome's northern frontier indicates that the cult was very popular among members of the legions stationed in those areas; and it is generally supposed that soldiers figured prominently in spreading Mithraism throughout the Roman Empire, as they were stationed in different areas and then settled down in various places when they retired from the army.

Although by this point Roman Imperial society was permeated by a belief in astrology, Mithraicists would have done their part in spreading such ideas, as suggested by the following inscription (*ILS* 4190) recording a dedication (probably a statue) made to Mithras by a public official in a small town of the Aequicolae in Latium: "to the Unconquered Mithras, Apronianus, the arkarius of the state, gave [this] gift, dedicated on June 25 when Maximus and Orfitus were consuls [172 A.D.], through the agency of the *pater* C. Arennius Reatinus."[32] The title *pater* was the highest of the seven Mithraic grades. Arennius Reatinus, therefore, would have been considered a kind of chief priest of Mithras in this small community. As is commonly the case in Mithraic dedications, Mithras is here given the epithet Invictus. As the performer of great exploits (especially the tauroctony), Mithras must have been regarded by his worshippers as a kind of Iranian Hercules; but on the other hand, he was also closely associated with, if not actually identified with Sol. Although Mithraic scenes show him sitting with Sol and enjoying a banquet of the meat from the sacrificed bull, most Mithraic inscriptions actually equate him with the sun, terming him *Mithras Invictus*, *Mithras Sol Invictus*, or *Mithras Deus Sol Invictus*. It is also highly significant that this dedication was made on June 25, the summer solstice in Roman times, when the sun was at the height of its power. As shown in the second part of this chapter, the Romans of the Republic did not associate solar worship with the solstices or equinoxes. The converse was true for sun worshippers of the second century.

Solar worship suddenly became very prominent in official Roman state cult, but in a radically different form, with the accession of Elagabalus in 218. Following the assassination of Caracalla in the previous year, the praetorian prefect Macrinus had become emperor, only to fall victim to assassination himself. At that point Julia Domna, the wife of Septimius Severus, who had founded the Severan dynasty, succeeded in reviving the Severan claim to imperial power by conspiring to place her fourteen-year-old grand-nephew on the throne. Before becoming emperor, Elagabalus had been serving as hereditary high priest to the Semitic sun god Elagabal of Emesa in the Roman province of Syria.[33] The city was the capital of a small principality ruled over by a hereditary line of priest-kings, who were not only rulers of Emesa, but were also the high priests of the community's solar divinity. Like Palmyra, situated about 100 miles to the east, Emesa benefitted from the traffic of luxury goods that passed through Syria to the Mediterranean coast for distribution throughout the Roman Empire. Moreover, since Syria from the beginning of the principate had been one of the militarized provinces, both trade and military service served to spread the cult of Elagabal to Rome, Italy, and the other provinces. The god's name was Latinized to Elagabalus and, like that of Mithras, was usually further amplified by the designation Sol Invictus. Inscriptions, as shown by Halsberghe (1972 45ff), chart the god's gradual advance during the second century, especially into the militarized provinces along the Rhine and Danube, but the cult achieved greater prominence with the accession of Septimius Severus to the imperial throne in 193 A.D., because Julia Domna was a member of the family that ruled Emesa. The city itself was also elevated to a more exalted status within the empire when Septimius divided Syria into two provinces and made Emesa capital of the northern one.[34]

As in earlier instances in which young men had become emperor (Gaius Caligula, Nero, and Commodus), Elagabalus' reign proved to be disastrous due to his immaturity and basic unfitness to rule.[35] Like Caligula, his behavior so alienated Roman public opinion that his reign lasted for only four years and ended in his assassination. Elagabalus shared his name with the god whom he served; and since the former had been born and raised to be Emesa's priest-king, his devotion to Elagabal was central to his self-identity. Consequently, when Elagabalus left Emesa to come to Rome, he brought with him the sacred black stone that had long been the god's representation in the temple of Emesa;[36] and On arriving in Rome the young emperor set about establishing his Syrian divinity in a new temple, the Elagaballium, on the Palatine next to his residence in the imperial palace (*Hist. Aug. Heliogab.* 3.4 and Herodian V.5.8). The emperor transplanted to Rome all the god's native rites, including his own use of make-up and the wearing of exotic silk clothes, thought to be quite appropriate for Roman women, but certainly not for a Roman man, and especially not for the Roman Emperor himself. Senators and knights were forced to be active participants in these rites. Elagabalus further alienated traditional Roman sentiment by marrying a Vestal Virgin, perhaps in imitation of the Near Eastern tradition of a sacred marriage, but also clearly intended to join together into one the two very different religious traditions of Rome and Emesa. The latter theme of religious fusion or syncretism is further illustrated by Elagabalus' bringing into the Palatine temple of his sun god all other religiously significant cults and artifacts, such as the black stone of Magna Mater, Vesta's fire, the *ancilia* of the *Salii*, and the Palladium.

Elagabalus built a second splendid shrine to his god on the eastern edge of the city (Herodian V.6.6–10) as part of an imperial residence that included a palace, (Sessorium), a race course (Circus Varianus), and an amphitheater (Amphitheatrum Castrense).[37] Herodian says that "at the height of summer"[38] there was a magnificent procession to this second temple from the one on the Palatine. It featured the black stone of Elagabalus being carried in a chariot to his other shrine, where public entertainments were held in the god's honor. In imitation of Augustus and Caligula, who had brought Egyptian obelisks to Rome to mark the midpoints of their race courses, Elagabalus did likewise for his own solar deity, placing on the *spina* of the Circus Varianus an obelisk measuring about 30 feet high, originally quarried in the time of Hadrian and erected in front of the tomb of his beloved Antinous at Antinoopolis in Egypt.[39]

Like Akhenaten 1500 years before, Elagabalus tried to bring about a religious revolution by elevating a solar divinity to supremacy; and like the Egyptian Pharaoh, the Roman Emperor failed, largely because his ideas were too radical to form a smooth and viable transition from well established religious traditions and current practices to his own idiosyncratic beliefs. When a second major religious innovation involving solar worship came about fifty years later, its author, the Emperor Aurelian, did not repeat these mistakes. Rather, he presented his subjects with a Hellenized and Romanized solar divinity that was totally in keeping with the mainstream of the religious culture of the day and hence guaranteed the cult's success.

Despite the failure of Elagabalus' religious revolution, its mere attempt testifies to the prominence achieved by solar worship in the Roman Empire at the beginning of the third century. As already noted, given the coincidence between the Day of the Sun of the planetary week with the Christians' Lord's Day, pagans often assumed that Christians were in fact sun worshippers. A curious Christian treatise, written in Latin and dating to 243, entitled *De Pascha Computus* (*Calculation of the Passion*), offers clear evidence that solar ideology was not simply being imposed upon Christianity from the outside by uninformed pagans, but as an important element in the *Zeitgeist* of the religious culture, it was being absorbed by Christians themselves and was having an impact upon how they constructed their theology. Although the work has not gone unnoticed among modern scholars in the history of early Christianity, the latter have generally failed to appreciate fully what it tells us about the degree to which Christianity was enmeshed in the wider religious culture of this period. Modern historians of Roman paganism seem to be entirely unaware of the work. At any rate, in attempting to arrive at the actual day upon which Jesus suffered on the cross, the author devises a complicated chronology, beginning with God's seven-day creation of the universe. He argues that the first day of creation must have occurred at the time of the vernal equinox on March 25, because on that day of the year light and darkness are equal in length, as they should have been when God began his work. Since the celestial bodies were brought into existence on the fourth day, they were created on March 28. Then after working his way through a long series of calculations, the author arrives at the conclusion that Jesus was also born on March 28, thus equating his birthday with the birthday of the sun in God's creation. These ideas, present in a Christian work, clearly demonstrate what we might term the solarization of religious thinking at this time. Equation between the Christian god and the sun is further shown by the fact that from the early third century onwards Christian writers occasionally termed Jesus *Sol Iustitiae*, the Sun of Righteousness (Strittmater 1942 615–617). The designation indicates that Christians countered the pagan equation of their god with the sun by conceptualizing God not as the fiery heavenly body that crossed the sky every day, but as a divinity that transcended all nature and had created all natural objects and beings.

During the 250's and 260's The Roman Empire teetered on the verge of total collapse, as the northern and eastern frontiers were simultaneously subjected to intense attacks.[40] The Germanic Goths, who had settled in the area around the Black Sea, crossed the lower Danube, plundered southeastern Europe, and even took to the sea in ships and carried out piratical raids throughout the Aegean, eastern Mediterranean, and Anatolia. During the 220's the Parthian dynasty that had ruled over Mesopotamia and Iran for more than three centuries was overthrown and replaced by the Sassanian dynasty of Persia, which initially adopted a much more aggressive posture toward Rome. As the Romans were distracted by these attacks, other Germanic tribes along the Rhine and Danube frontier carried out their own raids into Roman provinces. All these successful invasions of frontier areas spawned further internal instability through the outbreak of civil war, as different provinces, eager to secure themselves from outside attack, declared their own regional military commanders to be emperor. Because of the extraordinary chaos, modern scholars of the Roman Empire have termed the 49-year period, 235–284 A.D., the Military Anarchy, characterized by foreign invasions, civil conflict, and the rapid turn-over of occupants of the imperial throne.[41] During these years about sixty men laid claim to the imperial title, of whom only two died of natural causes. The Emperor Valerian (253–260) suffered the unique ignominy of not only being defeated at the hands of the Sassanians, but was even captured and died in captivity. This disaster was so demoralizing that his son, Gallienus (Emperor 260–268), had to contend not only with frontier wars against invaders, but also against Roman provinces that had seceded to mobilize their own defense under the command of military leaders who were themselves eager to lay claim to the Imperial purple.

One important religious consequence of the Empire's turmoil was the so-called first great persecution of the Christians, instituted by the Emperor Decius (249–251).[42] Since the time of Nero (Tac. *Ann.* XV.44) Christians had been persecuted, but only sporadically and locally, never systematically throughout the whole Empire.[43] In the words of Trajan's famous reply to Pliny on the subject (*Ep.* X.97) Christians were not to be sought out and punished, but if they came to the attention of government officials, then the matter had to be addressed. Christian monotheism forbade its true believers from participating in the polytheistic public festivals of the communities in which they lived. Their refusal to worship the gods whom everyone else worshipped laid them open to the charge of being atheists. Consequently, whenever some major crisis afflicted a community, such as a drought or pestilence, popular superstition could result in persecution, because by failing to believe in the traditional gods of the community Christians were suspected of having angered the gods, who were thought to be responsible for the crisis. Thus, forcing Christians to renounce their beliefs or eliminating them from the community was regarded as a proper way of regaining the goodwill of the gods (*pax deorum*) and thereby to bring the crisis to an end. As Tertullian wrote (*Apol.* 28.3), whenever disaster struck a community, the cry went up, "Christians to the lions! (*Christianos ad leones!*)." As a result, when the major disaster of the third century fell upon the whole Empire, many traditionalists viewed Rome's difficulties as stemming in part from the state's loss of divine approval, which in turn resulted from having so many non-believing Christians in their midst. Thus, the solution to recapturing divine favor was to force the non-believers into conformity and even to execute them if they did not relent. Decius issued a decree that required all inhabitants of the Empire to come before local officials to demonstrate their loyalty to the traditional gods by token forms of observance, such as pouring a libation, making a sacrificial offering, or eating the meat of a sacrificed animal. All those who complied were issued a signed document, attesting to their conformity.[44] Those who could not produce such a certificate were liable for punishment. In an inscription discovered at Cosa on the Etruscan coast in Italy (*AE* 1973 #235) Decius is styled "the restorer of the sacred rites (*restitutor sacrorum*)." Although the persecution ended with Decius' brief reign, it was revived a few years later under Valerian in a different guise. Rather than punishing and executing individuals, the state proceeded against Christianity as an institution by depriving churches of their properties and stripping any Christian members of the Roman senatorial and equestrian classes of their status; but this persecution too was called off by Valerian's son Gallienus, because he needed to mobilize all the Empire's energies in its struggle for mere survival.

Following Valerian's capture by the Persians, raids across the Rhine into Gaul by the Franks in the north and the Alamanni in the south resulted in Postumus being declared emperor in this region of the Empire, and its independent status persisted for fourteen years (260–274). At the same time, Persian attacks upon the Roman eastern provinces paved the way for the affluent Syrian city of Palmyra, first under Odenathus, then his wife Zenobia after the former's death, to emerge as a bulwark against the Sassanians.[45] Thus, the Roman Empire was threatening to break up into at least three parts. At this point in history, however, the Roman Empire gained a respite from the chaos by coming under the harsh, but effective rule of Aurelian.[46]

Though his reign lasted for only five years (270–275) and ended once again in an emperor's assassination, Aurelian succeeded in reuniting the Empire once again, thus richly earning one of his titles, Restorer of the World (*Restitutor Orbis*). His reign falls roughly into four parts. At its beginning he fought wars along the lower Danube area to drive out Goths, who had been ravaging the region for the past several years, after which he was forced to turn his attention to Italy to defeat and drive back invasions by the Juthungi. Having secured the central part of the Empire, he embarked upon the second phase of his reign by mounting a major expedition against Palmyra, which by now, under the rule of Zenobia, had extended its control over Syria, Judea, Roman Arabia (modern Jordan), and part of Egypt, and had also established a loose hegemony over much of Anatolia. While a Roman fleet recaptured Egypt, Aurelian proceeded overland with an army, crossed into Anatolia, reestablished Roman control, and then descended into Syria. After defeating the Palmyrene army at Antioch and capturing the city, he pursued the enemy to Emesa, where he won another victory. Zenobia and her forces then fell back to Palmyra, which the Romans proceeded to besiege. When Zenobia attempted to flee eastward to cross the Euphrates to find refuge among the Persians, she was captured, and Palmyra was persuaded to surrender. After settling affairs in the region, Aurelian headed back west, but *en route* Palmyra's revolt obliged him to turn about and to capture the city once again. It was sacked and left in ruins, never to arise again. The trade route across Asia that had made it so prosperous shifted to the north. Aurelian then returned to Italy and set into motion the third major military expedition of his reign, a campaign against Gaul, whose defeat ended its independence and brought it back into the Empire, after which Aurelian returned to Rome to celebrate in 274 his victories in the form of a traditional Roman triumph, but in 275 he was conspired against and murdered.

In addition to being important for the military and political history of the Roman Empire, Aurelian's reign proved to be quite significant in the history of Roman religion. In 274, following his reunification of the Empire, he had constructed in the Campus Martius a temple to Sol Invictus (Richardson 1992 363–364), remarked upon by later ancient writers for the splendor of its furnishings.[47] The Calendar of Furius Philocalus of 354 A.D. (Degrassi 1963 261) records December 25 as *N(atalis) Invicti* (= the birthday of the Unconquered), thus indicating that the shrine was dedicated on the winter solstice when the sun is at its most southern point in the sky before resuming its northward movement, hence serving quite naturally as the sun's annual birthday. The calendar further indicates that the temple's *dies natalis* was commemorated by thirty chariot races (*circenses missus XXX*). This celebration would have formed a welcome addition to the seven-day period of the Saturnalia (December 17–23), Rome's most joyous holiday season since Republican times, characterized by parties, banquets, and exchanges of gifts (see Books XIII–XIV of Martial). Aurelian clearly did not intend the shrine to be simply another temple to occupy space in a city already full of sacred edifices: for he established for the cult a new board of pontiffs, termed *pontifices Dei Solis* = pontiffs of the Sun God. They henceforth became part of the official religion of the Roman state alongside all the other public priests, including the traditional pontiffs, who were thereafter called *pontifices maiores* or *pontifices Vestae* in order to distinguish them from those of the sun god. In addition, there were games held every four years on October 19–22, probably to commemorate the anniversary of Aurelian's triumphal processions (Degrassi 1963 257 with Halsberghe 1972 16 and 18). Since the Greek word *agon* is used to describe these games, they were probably athletic, musical, and literary contests. It therefore seems obvious that unlike Elagabalus, Aurelian instituted a solar cult consonant with the traditional patterns of Greek and Roman religion.

But who was Aurelian's Sol Invictus? Our two chief ancient historical accounts, the *Historia Augusta* and Zosimus, imply that he was a god of Emesa or Palmyra respectively, and modern scholars have generally adopted one of these two views. In the *Historia Augusta*'s account of the battle of Emesa (25.3–25.6), when the Roman cavalry attack was flagging at a crucial point, a divine apparition manifested itself and rallied them on to victory. Immediately after the battle Aurelian entered the city and proceeded to the temple of the Sun, as if to discharge a vow made to the god. When the same divine apparition encountered him in the sanctuary, he established temples there in honor of the god, after which he had the one constructed in Rome. Then in 31.9 the *Historia Augusta* portrays Aurelian as having a priest summoned from Emesa in order to dedicate his temple of Sol Invictus in Rome. Consequently, some modern scholars have concluded that Aurelian's Sol Invictus was a modified form of Sol Elagabal of Emesa. Indeed, Halsberghe (1972 105ff and 136ff) has argued that despite the *damnatio memoriae* of the Emperor Elagabalus, the worship of his solar deity continued after his death and had become so generally accepted that it naturally became the model for Aurelian's own solar cult. Other scholars, however, have regarded Palmyra's worship of the Sun as a much more likely candidate (see Wissowa 1912 366–367 and Watson 1999 195ff). In his account of Aurelian's reign Zosimus says that when the temple of Sol Invictus in Rome was built, Aurelian adorned it with the riches plundered from Palmyra and also consecrated in it two of Palmyra's cult statues, one of Helios (Sol) and Belos (Semitic Bel). Given the brevity of Zosimus' account of this matter, his mention of the two cult statues from Palmyra appears to be emphatic, and the natural presumption might be that the historian intended to indicate that Aurelian's Sol Invictus was an adaptation of the solar god of Palmyra. Indeed, this interpretation would fit a well established pattern of Roman religious history. Since the days of the early Republic it had been customary for Roman generals to bring back to Rome the cults of conquered peoples in order to deprive the defeated of divine support and to enhance that of Rome. Aurelian therefore might have employed this formal religious procedure, known as *evocatio*, in summoning forth one or more gods of Palmyra to abandon the city and to come to Rome to receive worship in a magnificent temple in the capital of the Empire.[48]

Hijmans (1996 119ff) has recently opened up a third approach to this question of the identity of Aurelian's solar divinity. He has rejected both an Emesan and Palmyrene origin of the god and has argued instead that the solar deity was none other than the Greco-Roman Sol Invictus that had become increasingly popular during the past century or so. The account of the Historia Augusta need not be taken seriously, whereas Aurelian's inclusion of two cult statues from Palmyra in his temple in the Campus Martius, as recorded by Zosimus, may not have been anything more than supplementing the real cult statue of his temple with images of other divinities, a practice typical of the syncretic polytheism of the day. Thus, according to this interpretation, Aurelian's solar cult was both thoroughly traditional and in keeping with the contemporaneous religious trends. Indeed, the basic traditionalism of Aurelian's solar cult is further suggested by the evidence of his coinage.[49] His coin types are quite traditional in portraying divinities long associated with the protection, prosperity, and ancestry of the Roman people: Mars, Hercules, Victoria, Fortuna, Mercury, and Venus. One of the most common coins during the first part of Aurelian's reign shows Jupiter, the supreme god of the Roman state, handing a globe, symbolizing Imperial rule, to the emperor. Sol emerges prominently on the coins during the second half of the reign, after he had elevated Sol Invictus to preeminence by his temple in the Campus Martius. The coins portray Sol in traditional fashion, wearing a radiate crown, holding a whip for driving his four-horse chariot across the sky, and sometimes portrayed with the chariot. One very rare series, issued toward the end of the reign, shows Sol on the obverse with Aurelian at an altar on the reverse. This type was probably minted to commemorate Aurelian's official consecration of the temple of Sol Invictus, because the legend on the reverse is *Aurelianus Aug(ustus) cons.* with the last element plausibly expanded as *cons(ecravit)*. The legend on the obverse is *Sol Dominus Imperii Romani* = the Sun, Lord of the Roman Empire (Mattingly-Sydenham 1927 301 #319–322). This unusual title suggests that in part Aurelian's promotion of solar worship was designed to provide the diverse inhabitants of the Empire with a kind of religious unity, just as Aurelian's reign was most famous for restoring the Empire's political unity under the rule of a single emperor.

Thus, by the end of his reign Aurelian had succeeded in creating a solar cult that was thoroughly Roman and therefore easily incorporated into Rome's state religion with its own major temple as cult center, its own board of pontiffs, quadrennial games, and the cult's birthday on December 25, the winter solstice. Solar worship had always been common throughout the Empire, but it had also been rather heterogeneous, consisting of several separate religious traditions of Rome's multi-cultural population. In keeping with the religious syncretism of the day Aurelian's solar cult possessed the necessary latitude to be venerated by all sun worshippers and to accommodate their own particular beliefs and practices.

SOL AND CHRIST FROM CONSTANTINE TO THEODOSIUS

Following the assassination of Aurelian, instability characterized the tenure of the imperial throne for the next decade until Diocletian became sole emperor in 285, which ended the period of the military anarchy and ushered in what modern historians now term the period of late antiquity or the Later Roman Empire (284/5–395). viewed in retrospect, the age was the final period of relative peace and prosperity for the Empire. When Theodosius died in 395, leaving his two young sons, Honorius and Arcadius, as rulers in Italy and Constantinople respectively, the eastern and western halves of the Empire began their slow drifting apart. The process of separation was largely the result of the attacks upon the Rhine and Danube frontier by different Germanic tribes, even more intense than what they had been during the mid-third century. They eventually succeeded in dismembering the western half of the Empire by replacing its Roman provinces with a motley collection of Germanic kingdoms that spelled the end of Roman civilization proper and formed the basis for the early Medieval period of Western Europe. The eastern half of the Empire, of course, continued on and gradually evolved into the Byzantine State, but during late antiquity, before the two halves of the Empire went their separate ways, Christianity rose to prominence, and within less than a century made its way from a religion persecuted by the Roman state to one that fully became part of the government establishment and set about persecuting all religious rivals into obscurity. This is certainly not the place to try to narrate *in toto* the history of this period, but it is entirely proper, feasible, and quite necessary to examine a number of key episodes that will allow us to appreciate fully the religious and cultural context in which December 25 passed from being the birthday of Sol Invictus to that of Jesus.

As the events of the third century had shown, the defense of Rome's frontier had become so acute that emperors could no longer rule the Empire from Rome through their legates placed in command of military forces in frontier areas, but instead, the wars with Germanic tribes or the Persians often required the actual presence of the emperor. On the other hand, the frontier was sometimes threatened in different sectors simultaneously; and since the emperor could not be in more than one place at a time, the intensified pressure on the frontier system had resulted in invasions and civil wars as regional commanders took charge and then had themselves proclaimed emperors. In order to provide for the Empire's defense, as well as to eliminate internal instability due to such conflicts over succession to the imperial throne, Diocletian created the tetrarchy (rule of four), in which the Empire was divided into eastern and western halves with each governed by an Augustus and a

Caesar. This gave the Empire four, not one, emperor-like figures to deal with external threats.[50] The two Augusti were co-emperors, whereas the two Caesars, who served under them were junior partners, so to speak; and when an Augustus died or decided to go into retirement, the Caesar was to be automatically elevated to the senior status of Augustus, and a new Caesar would be chosen in his place. There was also an important religious element to this arrangement. As all emperors who had gone before them, the tetrarchs were eager to assert their special relationship with the gods, who had placed them in power, and whose agents on earth they were for watching over mankind. The four emperors advertised their connection to the divine by one pair claiming to be Jupiter's vice-regents, whereas the other did the same with respect to Hercules.

By 293 Diocletian's tetrarchy had come into being. He was Augustus of the eastern half of the Empire, whereas Maximian was Augustus for the western half. Galerius and Constantius Chlorus served Diocletian and Maximian respectively as Caesars. The eastern rulers were the Jovian pair, and the western were the Herculian. One decade later in 303, while presiding over a sacrifice and the inspection of the entrails by diviners, Diocletian became furious when he was persuaded that the ritual had been disturbed by the presence of Christians, who had been crossing themselves as an apotropaic sign for protection from the pollution or demonic nature of the pagan rites. He therefore decided to embark upon the second great persecution of the Christians. Thanks to two Christian writers of the early fourth century, Lactantius' and Eusebius, we are reasonably well informed about the interaction between Christians and the Roman state for the years 303–324. Lactantius, a Latin rhetorician and devout Christian, wrote a historical treatise on the persecution entitled *On the Deaths of the Persecutors (De Mortibus Persecutorum)*, in which he described the course of the persecution, God's vengeance upon the persecutors, and how Christianity emerged triumphant with Constantine as its champion. Eusebius, the bishop of Caesarea in Judea, wrote in ten books the first important history of Christianity (*Historia Ecclesiastica*), starting from the reign of Augustus and ending with Constantine's defeat of Licinius in 324 and emergence as sole ruler of the Empire. Books VIII–X cover the same ground as Lactantius' account.

According to Lactantius (12.1), following the incident involving the Christian disturbance of the sacrifice and divination, Diocletian formally began the persecution on February 23, the Terminalia in the Roman calendar. This day was obviously chosen quite deliberately in order to ensure the state's success in exterminating Christianity. The persecution lasted for eight years and seems to have been prosecuted much more aggressively in the east than in the west. Lactantius (15.7) asserts that Constantine's father did little more in his provinces other than demolishing some churches and having sacred texts destroyed. The persecution was ended in 311 by Galerius, who had replaced Diocletian as eastern Augustus. In the following year Constantine, who was Caesar in the west, became the first imperial ruler to embrace Christianity; and in 313 he and his imperial colleague Licinius issued the edict of Milan that proclaimed religious toleration throughout the Empire, thus allowing Christianity to be practiced without fear of persecution. But before delving further into these events and their relevance to the subject of this chapter, a few sentences must be devoted to sketching the demise of Diocletian's tetrarchy.

Despite the simplicity and logic of the tetrarchy, the scheme, as a system to ensure a peaceful imperial succession, failed even within Diocletian's own lifetime, because it did not or could not fully take into account two important and interrelated factors: (1) the natural tendency within a hierarchical society for the hereditary principle to be axiomatic in determining the succession; and (2) the overweening ambition of members of imperial families, who regarded themselves as entitled to imperial power. When the two Augusti, Diocletian and Maximian, went into voluntary retirement on May 1 of 305, they were succeeded by their Caesars, into whose places were promoted Severus in the west and Maximinus Daia in the east. But within the space of a year or so this pattern of smooth transition was thrown into chaos by two usurpations.[51] When Constantius Chlorus died at Eburacum (York) in Britain in July of 306, his son Constantine, without waiting to be appointed Caesar, stepped into the vacated position and was hailed Augustus, not Caesar, by the army at hand. Three months later in Rome Maximian's son, Maxentius, also laid claim to imperial power, so that the western half of the Empire had three persons vying for power: Severus, Constantine, and Maxentius. When Severus invaded Italy to eliminate Maxentius, he was defeated, taken prisoner, and died in captivity, thus leaving Maxentius in control of Italy and North Africa with Spain, Gaul, and Britain under Constantine. Eventually, as might be expected, civil war broke out between the latter two, as each strove to become master of the entire western half of the Empire. The struggle culminated in 312 with Constantine's invasion of Italy and march on Rome; and it ended with the battle at the Mulvian Bridge just north of Rome, in which Maxentius was defeated and killed, whereupon Constantine became the unchallenged emperor of the western provinces.

According to Christian writers, it was during this campaign that Constantine had a profound religious experience that brought about his conversion to Christianity. We possess three rather different accounts of this crucial event. The most elaborate version occurs in Eusebius' Life of Constantine, written 25 years or so after the event, between the deaths of Constantine in 337 and of Eusebius in 339. According to this account (I.27–32 and 37–40), the experience occurred as Constantine was setting out from Gaul to attack Maxentius' forces in Italy. In order to obtain divine assistance in this risky enterprise, Constantine prayed to the god worshipped by his father, asking him to reveal himself.[52] Then in the early afternoon both Constantine and his army saw a cross above the sun with the words "in this, conquer!" During the ensuing night Jesus appeared to Constantine in a dream and showed him what kind of military standard he should have constructed as the Christian emblem for securing victory. Upon awakening and learning through inquiries that this divinity was the monotheistic god of the Christians, Constantine resolved to embrace Christianity and instructed craftsmen to form a standard in the shape of a cross, over whose intersection was placed a bejeweled gold wreath encircling the famous Chi-Rho symbol, an X vertically bisected by a P, because in Greek these were the first two letters of Christ. Armed with this Christian standard, later termed the *labarum*, Constantine proceeded against Maxentius' forces, defeated three armies in northern Italy, marched upon Rome, vanquished and killed Maxentius himself, and entered Rome in triumph. In commemoration of his victory with the aid of the Christian god there was erected a colossal statue of the emperor with a large cross held in his hand.

In his *Ecclesiastical History*, written a generation earlier and shortly after the battle of the Mulvian Bridge, Eusebius (IX.9.1–14) tells a very different story. Constantine is again the historical agent of the Christian god in destroying Maxentius, but he is not the recipient of any astonishing heavenly phenomenon. He simply prays to the Christian god and Jesus, defeats all the forces that stand in his path, kills Maxentius, enters Rome in triumph, and has a statue of himself holding a cross set up to commemorate his victory. Throughout the account God manifests himself only through the normal events of human affairs. Eusebius, however, attempting to endow his account with suitable divine trappings, likens Maxentius' destruction to that of Pharaoh In *Exodus* in his pursuit of the Hebrews.

In Lactantius' version, also written shortly after the end of the civil war, Constantine's momentous religious experience occurs, not in Gaul at the beginning of the expedition to Italy, but during the night before the final battle with Maxentius at the Mulvian Bridge. Lactantius (44.5–44.6) simply says: "Constantine was advised in a dream to mark the heavenly sign of God on his shields and to engage in battle. He did as he was commanded. He marked Christ upon the shields, the letter X pierced through with the very top bent around. Armed with this sign, the army took up its weaponry."[53] Lactantius then proceeds to describe the battle in terms similar to those of Eusebius with additional details concerning Maxentius' behavior in Rome before the battle.

Needless to say, this event with its different versions has prompted considerable modern debate. What reasons led Constantine to become a Christian? Did he in fact become a Christian? If so, to what extent? Did he have a vision or dream that caused him to embrace Christianity?[54] As MacMullen has shown so convincingly (1968), the culture of late antiquity, both Christian and Pagan, was thoroughly permeated with a firm belief in the supernatural that regularly manifested itself in human affairs. In opposing some modern views of Constantine as a shrewd and calculating *Realpolitiker* in siding with Christianity, Jones (1962 73) was certainly on target when he observed, "to be a rationalist in that age Constantine would have been an intellectual prodigy." Indeed, Lactantius' account is completely in keeping with centuries of Roman tradition in which generals, about to engage an enemy in an important battle, called upon divine assistance and promised to honor the deity with a temple or magnificent dedication in the event of victory. In addition, dreams were the most common means by which people in the ancient Greco-Roman world received communications from divinities. Thus, Lactantius' account is entirely consistent with the well established religious culture of the day. It therefore could actually be historically accurate; or alternatively, its plausibility and verisimilitude have been generated out of the normal assumptions of the prevailing religious culture.

On the other hand, despite its fabulous nature, Jones (1962 85–86) offered a plausible explanation for the celestial phenomenon set forth in Eusebius' later account: namely, the so-called halo effect in which ice crystals in the upper atmosphere reflect and refract sunlight so as to produce a double image of the sun, to make the sun encircled by rings, or to frame it in some other shape, such as a cross. In fact, such phenomena are found in the prodigy lists recorded in Republican times.[55] Eusebius reinforced his later version by stressing that he had heard this story from Constantine himself, who took an oath as a guarantee of his veracity. In any case, it is quite clear that as children of their culture, Lactantius, Eusebius, Constantine, and all his soldiers must have had no difficulty in believing that they and the world around them were constantly being impinged upon and influenced by supernatural forces.

Unraveling the truth behind Constantine's supposed Christian vision or dream is further complicated by the report of an earlier pagan vision that he experienced. While in his Imperial capital of Trier in Gaul in 310, Constantine was addressed by an unknown Latin rhetorician, who delivered a speech whose text has come down to us as the sixth in a collection of similar imperial addresses of this period known as the *Panegyrici Latini*.[56] Constantine had recently marched south to suppress a rebellion led by Maximian, Diocletian's Augustan partner, who had emerged from retirement once again to usurp power; but this time, when his attempt ended in failure, he committed suicide. In Constantine's absence from the Rhine frontier, the Germanic tribes threatened invasion, thus obliging the emperor to turn about on a forced march. Hardly had the march begun, when news was received that all was well. Constantine then took advantage of the respite to turn aside to visit a temple of Apollo to discharge vows that he had made for his success. We know of this event only from the following passage of the Gallic orator (*Pan. Lat.* VI.21.3–7):

> On the day after that report had been received [of a possible Germanic invasion across the Rhine], and after you had undertaken the labor of a double march, you learned that all the waves had subsided, and there had returned all the tranquility that you had left behind. Fortuna herself so ordained this that the felicity of your affairs might advise you to render to the immortal gods what you had vowed at that place whither you had detoured to the most beautiful temple in the entire world, nay! rather, to

> the god manifest, as you saw: 4 for you saw, Constantine, I believe, your Apollo, accompanied by Victoria, offering to you laurel crowns that each bore the presage of thirty years. This is the number of human ages that are certainly due to you, beyond the old age of the Pylian. 5 Nay! Why do I say, "I believe?" You saw and recognized yourself in the specter of *him* to whom were owed the regions of the world as the divine verses of the poets have sung. 6 This now, I think, has finally happened, since you, like that one, are an emperor youthful, joyous, health-bringing, and most handsome. 7 You have therefore deservedly honored that most august shrine with the greatest offerings.[57]

The Gallic orator was obviously setting forth ideas already publicized by Constantine, who wished to have them broadcast as much as possible, because they were important elements in his imperial ideology. The phrase *praesentem deum* (god manifest) demonstrates beyond doubt that Constantine wanted it to be known that he had had a personal encounter with a divinity; and the orator's *Apollinem tuum* (your Apollo) informs us of this god's identity, as well as stressing the special relationship between Apollo and the emperor. As Rodgers has argued (1980 264), since Constantine had just returned from quelling the rebellion led by the Herculian Maximian, it is likely that Constantine was now eager to repudiate his Herculian ties and construct a new divine association. The conjecture that Constantine at this time was formulating a new imperial ideology by connecting himself with Sol receives further support from his bronze coinage, which from 310 onwards is dominated by representations of this divinity. On the other hand, by exploring similarities between this passage and Vergil's *Aeneid*, Rodgers has further argued that the ambiguous allusion in 21.5–6 is not to Apollo, but to Augustus, whom Constantine was now adopting as his imperial model. Be that as it may, the wording in 21.5 clearly recalls Vergil's *Fourth Eclogue*, foretelling the return of the golden age under the rule of Apollo. In this regard it is most noteworthy that the 110-year Augustan *saeculum*, marked by the *Ludi Saeculares* of Septimius Severus in 204, was drawing to an end. Vergil's *Fourth Eclogue* is therefore likely to have been in the minds of at least some well informed persons of Constantine's court, so that the emperor might have entertained the notion of himself as a latter-day Augustus for the new *saeculum*.

Scholars such as Grégoire (1930) and Hatt (1950) have rightly focused our attention on the possible symbolism of the laurel wreaths in Constantine's alleged pagan vision and their connection with the famous *labarum* described by Eusebius in his *Life of Constantine*. These scholars are doubtless correct in thinking that the wreaths promising Constantine a reign of thirty years were circles circumscribing X's, the latter obviously representing the number ten, so that three such symbols would signify thirty years. It had long been customary for Roman emperors to celebrate the tenth and twentieth years of their reigns, at which time vows (*vota publica*) were always made for another ten years. When these commemorations and vows were memorialized by being inscribed on stone, they always took the abbreviated form of X, XX, or XXX. Moreover, Grégoire noted that the upper half of the famous Chi-Rho monogram was the same as a ligature of a V and P representing *vota publica*. According to Eusebius, the *labarum* was a cross with a gold bejeweled wreath attached to the intersection, inside of which was the Chi-Rho monogram. Removing the Rho simply produces a circle circumscribing an X, which, as Grégoire and Hatt argued, was initially what Constantine had allegedly seen in his pagan vision. It therefore seems likely that the Christian *labarum* did not arise *ex nihilo*, but was, if anything, simply a slightly modified version of a banner or military standard employed by Constantine in Gaul before he ever contemplated invading Italy to overthrow Maxentius. Since Constantine's early years as a ruler were spent in the Celtic provinces of Britain and Gaul, whose adult male inhabitants must have composed the majority of his army, Hatt postulated that the *labarum*'s origin should be sought in Celtic symbols. According to Webster (1986 130) a circle circumscribing an X was a common Celtic religious symbol representing the sun. It is found on pottery and cultic artifacts. Its shape probably was intended to symbolize a wheel, whose rapid rotary motion suggested the swiftness of the sun in moving across the sky, like the spoked wheels of the Sun's chariot standard in Greco-Roman art. Thus, the Gallic orator's mention of "your Apollo" and the wreaths circumscribing X's lead us back into the realm of solar worship.[58]

Sol Invictus figured very prominently on the coins of the tetrarchy, thus making it very likely that Constantine was in fact a devotee of the Sun. Imperial coins during the tetrarchic period bore numerous types of the deity with legends such as *Soli Comiti Augusti* (= to Sol, the Emperor's Companion), *Soli Invicto*, *Soli Invicto Conservatori*, *Soli Invicto Aeterno*, and *Soli Invicto Comiti*.[59] Even after Constantine's conversion in 312 Sol Invictus continued to be shown on his coins for another eight years (Bruun 1958, cf. Alföldi 1948 57–59). In 308 Diocletian came out of retirement briefly to confer at Carnuntum (east of modern Vienna on the upper Danube) with the tetrarchs in order to eliminate the turmoil created by the recent usurpations. Galerius and Licinius were confirmed as the Augusti of the east and west respectively with Maximinus Daia and Constantine serving as their Caesars. The conference was commemorated by a sculpture group portraying the four rulers standing together in two pairs with one of their hands on the other's shoulder to indicate solidarity, but with their other hand gripping their swords to display their sternness. At this same time the tetrarchs were responsible for the restoration of a Mithraeum at Carnuntum, whose inscription (*ILS* 659) reads as follows: "to the god Sol Invictus Mithras, supporter of their Empire, the most religious Jovian and Herculian Augusti and Caesars restored [this] shrine."[60] As the inscription indicates, by now Mithras had been fully merged with Sol Invictus; and as shown by inscriptions testifying to the building or repair of Mithraea, the composite deity was very popular among the soldiers serving along the Rhine and Danube frontier.[61] Another inscription from Mauretania Sitifensis (*CIL* VIII. 8712) records the construction of a shrine to Sol "for the welfare (*pro salute*)" of Constantine and Licinius, who are styled "the most unconquered and forever Emperors (*invictissimi, semper Augusti*)." The structure was erected by the equestrian governor of the province, who, we assume, would not have done so if he had thought that it was inconsistent with the emperors' religious attitudes.

Since it was well known that Christians worshipped their god on Sunday, Constantine and many others must have been initially attracted to the faith through its presumed connection with solar worship. Moreover, as devotees of the Unconquered Sun filtered into the Christian community, their beliefs are likely to have affected their newly adopted religion. This is suggested by the fact that Lactantius (2.1) places Jesus' resurrection on March 25, the vernal equinox, which was also the Hilaria in the cult of Magna Mater, marking the resurrection of Attis. Lactantius gives this date for Jesus rising from the dead in such a matter-of-fact way that it must have been well established belief. As one of the early Christian polemicists against all forms of paganism, including worship of the sun, Lactantius as our source for this dating of the resurrection is even more remarkable. He betrays no uncertainty, hesitation, or qualms about the matter, which certainly means that the association between the vernal equinox and the resurrection was so well established through religious syncretism that by Lactantius' day no one gave the matter much serious thought.[62]

But to return to Constantine, as an imperial prince, his youth must have been fully centered around his training to become a ruler, which above all, given the circumstances of the times, would have meant a practical education in military affairs. As all three Christian accounts of Constantine's conversion make clear, his paramount concern in 312 was to achieve military victory over Maxentius and his forces, and resorting to the divine was simply part of his overall strategy. Thus, the following assessment by Alföldi of Constantine's conversion (1948 19–21) would appear to be entirely accurate:

> It is clear from what we have been saying that the ideas and inferences associated with the vision of Constantine were nothing but abortions of the excitable religious fancy of the late Roman Empire. Constantine was prevented from breaking away from this debased form of religion by his inferior education. Some scholars, indeed, suppose him to have grown up in the atmosphere of the palace and to have realized, from the first, the importance of thorough and individual education of the personality. But such was not the case. The rough, good-tempered Illyrian soldier in whose house Constantine grew up had laid little stress on letting his son have a good education; what he wanted above all was to make him a good general and efficient administrator. At the most, we may allow that Constantine acquired in the air of a palace decent principles of respect for culture, and that this spurred him on to let his own sons have a really thorough classical education. For all his high birth Constantine was certainly not well educated; he was *litteris minus instructus*, to quote the

> words of a reliable authority. A letter of his, it is true, survives, written in beautifully smooth and resonant phrases, which show off the perfect literary education of its composer; but that only makes it the plainer that the brilliant stylist was not the Emperor, who signed it, but a man of letters in the Imperial chancellory. What Constantine actually composed himself—letters discussing Christian doctrine and the like—are ponderous and wordy, long-winded compilations. A similar clumsiness and unbridled passion is revealed in his legislation. If Constantine's acquaintance with worldly literature was slight, his knowledge of the Bible was equally weak. The subtle speculations of theology were a closed book to him. A distinguished modern Church historian has branded his letter to Bishops Alexander and Arius as a forgery, because he was scared by the pieces of ' childish silliness' that occur in it. But such primitive ideas characterize the whole of Constantine's religious writings and betray thereby their imperial composition. . . . It is just these written declarations of Constantine that betray that the dogmatic foundations of Christianity remained a mystery to him. What really gripped this son of an age of decadence, sunk in superstition and mysticism, was not the refined theological system of the Church, not the lofty moral teaching of the New Testament, but its unbounded faith in the limitless power of Christ. From it he expected the prosperity of his Empire in peace and its victory in war. So true is it that power was for him the deciding factor, that he did not attach himself to Christ until Christ had fulfilled His promise to lend him His aid. Here we see a survival of that way of thinking that characterizes ancient paganism: *votum solvit libens merito*–to quote the expression of thousands of dedicatory inscriptions.

Constantine's excitement in his newly found source of divine power emerges from an event of early 313 described by Lactantius (46–47) as analyzed by Jones (1962 77–80) in his characteristically masterful way:

> On 30th April Maximin deployed his troops for battle. Licinius, despite the fact that he was outnumbered by more than two to one, accepted the challenge. For he did not rely on human resources alone. As his troops came into line they grounded their shields, removed their helmets, and raising their arms to the sky, recited in unison, their officers dictating the words, the following prayer: "Highest God, we beseech thee, Holy God, we beseech thee; to Thee we commend all justice, to Thee we commend our safety, to Thee we commend our Empire. Through Thee we live, through Thee we are victorious and fortunate. Highest, Holy God, hear our prayers: we stretch out our arms to Thee; hear us, Holy, Highest God." The battle was swift and decisive. Maximin, flinging off his imperial robes and disguising himself as a slave, fled posthaste for the straits. . . . [After quoting part of the Edict of Milan agreed upon a few months earlier by Constantine and Licinius in guaranteeing universal

> religious toleration, Jones continues] From these events it is possible to reconstruct what had passed at Milan. Constantine and Licinius had agreed on a common policy towards the Christians: the property of the Church was to be restored and full and untrammelled liberty of worship permitted. Licinius' edict bears signs, in its laborious insistence that both Christians and others were to enjoy toleration, of being a compromise, and there can be little doubt in which direction either emperor was pulling. Constantine had already in his own dominions gone further than mere restitution and toleration: it must then have been Licinius who insisted on a strict impartiality. It would also appear that Constantine had urged Licinius, in his forthcoming campaign against Maximin, to place his armies under the protection of that heavenly power which had granted his own armies victory over Maxentius. This advice Licinius apparently accepted with reservation. He did not adopt the sign under which Constantine's men had fought, and he drafted a form of prayer which, while it should be acceptable to the heavenly power, could give no offence to any other god.

We need not doubt that Licinius and many of his soldiers equated this supreme god, recently adopted by Constantine, with Sol Invictus Mithras. Indeed, Licinius' devotion to Sol Invictus is revealed by an inscription (*ILS* 8940) recording the consecration of a statue to *Deus Sanctus Sol*, who on Licinius' orders was to be honored annually with incense, candles, and libations on November 18, the day of the statue's consecration.

In 321 Constantine issued an imperial decree that established Sunday as a day of rest: "All judges, townsfolk, and shops of all crafts are to rest on the venerable Day of the Sun. Yet, those situated in the countryside for the cultivation of the fields may work freely and openly, because it often happens that there is no better day for entrusting the grain to the furrows or the vines to the ditches, lest the suitability granted by heavenly providence be lost from the opportunity of the moment."[63] In view of the emperor's probable earlier solar worship this decree takes on greater significance. The decree is confirmed and further elucidated by Eusebius' *Life of Constantine* IV.18–20, according to which not only was Sunday a day of rest and worship for Christian members of the army and imperial palace, but soldiers who were not Christian were obliged to march out into an open field, where they raised their faces and hands skyward and uttered a prayer to the supreme deity composed by Constantine himself. Since under such circumstances the sky would have been dominated by the sun, we may plausibly suppose that many soldiers would have equated Constantine's supreme divinity with the well-known pagan Sol Invictus Mithras.

Another solar association with Constantine was the emperor's bronze statue erected on top of a porphyry column in the center of the Forum of his new city, Constantinople, formally dedicated on May 11 of 330, and which is generally thought to have been a thoroughly Christian community. The statue, however, is likely to have been one of Apollo recycled as that of the emperor, bearing seven rays on its head and mounted to face east toward the rising sun.[64] Since twenty-five years elapsed between the battle of the Mulvian Bridge and Constantine's death, it is likely that during these years Constantine's religious thinking and behavior underwent a slow evolution from traditional paganism toward Christianity, as the result of his interaction with devout Christians well versed in Biblical literature and theology. In any case, the surviving evidence makes it abundantly clear that pagan solar worship had played an important role in bringing about Constantine's conversion to Christianity.

One last item regarding Constantine and the sun must be touched upon, and it involves another Egyptian obelisk. In 357 there was erected in Rome the largest of the Egyptian obelisks, measuring about 100 feet tall and originally quarried during the reign of Egypt's greatest warrior pharaoh, Tuthmosis III of the eighteenth dynasty during the fifteenth century B.C. (now in the Piazza S. Giovanni in Laterano). The massive monument was placed upon the *spina* of the Circus Maximus to join Augustus' obelisk. The four faces of the base were inscribed with four sets of six hexameter verses that narrated the story of the obelisk's journey from Egypt to Rome (*ILS* 736). According to these lines, Constantine had planned to bring the shaft from Egypt to Constantinople; but death prevented him from carrying out his plan. Some time later, his son Constantius, who had succeeded him as emperor, had the monument conveyed by ship to Rome, but before it could be erected, the western half of the Roman Empire revolted under the usurper Nagnentius. Finally, after Constantius visited Rome for his first and only time in 357 to celebrate a triumph, the obelisk was erected in the Circus Maximus, around whose track the emperor must have ridden in his triumphal chariot. The last two lines of the inscription reads: "The victor, triumphing and favoring the city, establishes the lofty trophy and gift of the emperor and adorns it with his triumphs."[65]

The inscription terms this monument a lofty trophy (*sublime tropaeum*). The term *tropaeum* (Greek *tropaion*) had its origin in Greek hoplite warfare of the archaic period and signified the place where the defeated army turned and began to flee from the victors (Greek *trope* = 'turning'). To commemorate their achievement, the victors erected a monument at or near this place and usually had it adorned with weapons and armor taken from the defeated. By the mid-fourth century of our era *tropaeum* was regularly used by Christian writers to refer to the cross, because to them it signified Jesus' victory over death and over sin for all mankind. Given the well-known association of the Circus Maximus with Sol, Constantius' erection of this obelisk to join that of Augustus, which was itself dedicated to Sol, raises the strong likelihood that the Christian Constantius intended this obelisk to honor the Christian god, whom he associated with the sun. Strange as this may seem, it is important to realize that it was during these very years that December 25, the birthday of Sol Invictus, was rapidly gaining acceptance as the birthday of

Jesus. Constantine must have intended to bring this shaft to Constantinople to be placed in the Hippodrome in imitation of Augustus' obelisk in the Circus Maximus. Thus, the same solar associations posited here with respect to Constantius should apply as well to Constantine.

The last great historian of the ancient world, Ammianus Marcellinus, records the erection of this obelisk in his narrative (XVII.4). The passage is quite remarkable. Not only does Ammianus treat the reader to an antiquarian digression on Egyptian obelisks, but he quotes extensively from a Greek translation made by a certain Hermation of the hieroglyphics carved on this monument. The quoted Greek text shows very clearly that the obelisk was dedicated to the Egyptian sun god, who in turn bestowed his grace upon Egypt's ruler. Why did Ammianus trouble himself to quote from this Greek translation? A few chapters earlier (XVI.10) Ammianus describes Constantius' visit to Rome, and he does so in the most scathing manner, stressing that Constantius had conquered no foreign enemy to justify a triumph in the traditional fashion. It is therefore quite possible that the historian's detailed treatment of the obelisk was similarly intended to criticize Constantius, but unlike his overt castigation of the emperor for his shortcomings in war in XVI.10, Ammianus has constructed his narrative in XVII.4 so as to criticize in a veiled manner Constantius' Christian piety by demonstrating to his readers that the erection of the obelisk confused Christian theology and pagan solar worship. It was one thing for a pagan historian to find fault with a deceased Christian emperor for his political or military achievements or lack thereof, but it was far more dangerous to impugn his Christian piety.

This last point serves to remind us that from 312 onwards the line between Christians and pagans became increasingly well defined to the detriment of the latter. As the fourth century advanced, there gradually came into being a carrot-and-stick system of incentives and deterrents designed to force people away from paganism and to drive them toward Christianity.[66] For example, under Constantine all members of the Christian clergy were granted complete exemption from taxes and liturgies, which during this period could cripple the fortunes of local elites.[67] It provided a powerful financial incentive for those of comfortable means to preserve their fortunes by signing up with Christianity. Thus began a trend that henceforth bedeviled the Church for centuries: persons who became priests for financial reasons, not because of a genuine religious calling. From Constantine onwards pagan temples began to be stripped of their wealth, which was often recycled along with large sums of public money to the greater glory of the Christian god in the form of lavishly adorned churches. Title 10 of Book XVI of the *Theodosian Code* records 25 imperial pronouncements by Christian emperors against various aspects of paganism, dating from Constantine in 321 to Theodosius II and Valentinian III in 435. These decrees present a gloomy picture of pagans marginalized and bullied by the power of Christian emperors. The official language characterizes paganism as superstition at best and as demonic at worst. From Constantius onwards pagan temples were closed, and all forms of sacrifice were forbidden, often under pain of death for disobedience. The religious policy of the Christian emperors, however, up until the last quarter of the fourth century was not uniform and unvarying, but rather, it vacillated between intolerance and suppression, on the one hand, and grudging toleration and bare coexistence on the other (Barnes 1989). Although paganism continued to exist, it was henceforth under continual assault; and many pagans must have been cowed by the power of the state, as the Christianization of the Empire advanced. The only interruption of the latter trend came with the brief one-and-a-half year reign of Julian (November 3 361 to June 26 363), during which the last pagan to sit on the imperial throne reopened the temples, reinstituted sacrifices, and attempted to revitalize paganism and to halt the growth of Christianity.[68] The religious history of this period is epitomized by a Greek taurobolic inscription from Rome, dating to Julian's reign, which celebrates the end of 28 years of night, i.e. the preceding generation during which the taurobolium had apparently been outlawed (Rose 1923). As indicated by Julian's *Fourth Oration to the Sun*, a highly syncretic form of solar worship, pantheistic and henotheistic, was part of the emperor's religious program in having a strong pagan candidate to rival the monotheistic god of the Christians.[69]

After Julian's death the suppression of paganism resumed and accelerated from the 370's onwards. Ammianus (XXVIII.1) describes how the city of Rome c.370 was thrown into a state of fear when the man in charge of the grain supply carried out a series of prosecutions involving allegations of sacrifices and magic, resulting in the wrongful executions of several high ranking persons. Shortly thereafter (Amm. Marc. XXIX.1–2) a similar, but much greater storm of charges and executions disturbed the imperial court at Antioch, stemming from a few individuals using a kind of ouija board to consult divine powers (certainly Apollo) concerning the identity of the next emperor. Following numerous executions of prominent people, a general campaign against all supposed forms of sorcery was waged throughout the eastern provinces. Since its casualties included the most eminent pagan philosophers, it succeeded in decapitating Greek paganism. Ammianus, who lived through these events, likened the atmosphere to that at the court of Dionysius, the famous tyrant of Syracuse, and to the story of Damocles, over whose head was hung a sword by a single horse hair. Besides the tortures, executions, imprisonments, and confiscations of property, countless books of Greek religion and philosophy were burned, sometimes by the authorities, but also by their owners, so as to prevent charges of sorcery being brought against the books' owners. Henceforth it was not uncommon for local Christian priests and monks to lead mobs of vigilantes in vandalizing or destroying some pagan shrine (see Libanius XXX). The Mithraeum below the Church of Santa Prisca on the Aventine in Rome is only one of many such casualties revealed by modern archaeology.[70]

Christian suppression of paganism intensified further during the closing two decades of the fourth century. Of the 25 imperial pronouncements against paganism recorded in Title 10 of Book XVI of the *Theodosian Code*, twelve (7–18) date to this period with item 12 (August 392) being the most comprehensive and universal in outlawing all forms of paganism and even establishing penalties for officials who failed to enforce the measures. In 383 the young Emperor Gratian, somewhat bookish, very devout, and heavily influenced by Ambrose, bishop of Milan and staunchly anti-pagan, became the first Christian emperor to repudiate the title of *pontifex maximus*, which had been among all emperors' titulature since the time of Augustus.[71] Gratian also cut off all public funds for activities traditionally associated with Rome's pagan state religion. The epigraphic record in the form of the taurobolic altars unearthed beneath the facade of St. Peter's in the Vatican show that a substantial number of the senatorial class in Rome at this time responded to the Imperial government's suppression of paganism by holding multiple pagan priestly offices and proudly advertising them in inscriptions (Matthews 1973). Then in August of 392, when Valentinian II, Emperor of the West, suddenly died without having made plans for a successor, this staunchly pagan group of Roman senators seized the opportunity to fight back against Christian religious intolerance. Eugenius now became emperor with their support, and for the two years of his reign paganism in Rome enjoyed its last hurrah![72]

What had been a culture war for the past eighty years now culminated in a real war. Paganism's last stand ended when Theodosius, Christian Emperor of the East, marched westward and encountered Eugenius and his forces. As the climactic battle approached, the pagans' confidence was buoyed up by a prophecy that Christianity's time was about to end, having existed for 365 years, a year of years (Aug. *Civ. Dei* XVIII.53). The decisive battle between Eugenius and Theodosius occurred at the Frigidus River (modern Vipava) that flows into the Adriatic at the head of the Gulf on the modern border of Italy and Slovenia. After a very hard fought battle on September 5 with heavy casualties sustained on both sides, Theodosius achieved victory by mounting a surprise attack early on the following day (Zosimus IV.58 and Oros. VII.35 with Seeck-Veith 1913 and Cameron 2011 93ff). His victory was greatly facilitated by a powerful wind that rendered the missiles of Eugenius' army ineffective (Aug. *Civ. Dei* V.26). Of course, nature's role in Theodosius' victory was hailed by Christians as divine intervention and as proof of the superiority of their deity. Eugenius was killed, and Theodosius entered Italy. After reversing Eugenius' policies and settling affairs generally, Theodosius returned to the East having left his son Honorius in Italy to be its new junior emperor. A few months later (January 17, 395) Theodosius was dead, and the two halves of the Roman Empire, now under the rule of his two sons, began their slow drifting apart to two very different historical fates, but by then Christianity had established itself as supreme in both the East and the West. Nevertheless, paganism continued to survive for centuries in two forms: intellectual paganism among the educated elite (i.e., adherence to the ancient traditions of paganism without the physical trappings of temples, sacrifices, and public ceremonies), and popular traditional pagan rites observed by the common folk in towns and rural districts, prompting emperors after Theodosius to issue decrees prohibiting pagan rites. In fact, many pagan religious traditions were so deeply woven into the fabric of society that they died out only when they were modified and absorbed into Christianity.[73]

DECEMBER 25, JESUS' BIRTHDAY

Having erected this elaborate historical backdrop and having scripted the plot of this most interesting tale, we may now bring onto the stage for a final curtain call December 25 as Jesus' birthday. As already noted, for this day the Calendar of Furius Philocalus of 354 A.D. has the mere two-word entry *N(atalis) Invicti* (Degrassi 1963 261). In the absence of other evidence, we might suppose that it was simply intended to note the anniversary for the dedication of Aurelian's temple to Sol Invictus, but another document, *Depositio Martyrum*, which has come down in the manuscript tradition along with the calendar, begins its list of Christian Saints days to be commemorated with the words: "*VIII Kal. Ian. natus Christus in Bethleem Iudeae* = On December 25 Christ was born in Bethlehem of Judea." This brief item clearly indicates that in 354 A.D. in Rome December 25 was regarded as Jesus' birthday, and it constitutes our earliest evidence for such a belief.[74]

As already established at the end of the first section of this chapter, by c.200 A.D. January 6 was widely regarded by Christians of the Greek East as the day of Jesus' birth; and as the result of other scattered bits of information, it appears that by c.300 Jesus' birth was being celebrated by Christians in the eastern provinces and perhaps in the western ones under the name of Epiphany. According to Ammianus Marcellinus (XXI.2.5), at the beginning of 361 Julian, soon to be Rome's last pagan emperor, observed Epiphany in Gaul. From the middle of the fourth century onwards this pattern began to change, as December 25 was taken over by eastern communities of Christians to replace January 6. Although we possess some data attesting to the eastward diffusion of December 25, it is rather meager, thus leaving many important questions unanswered. This situation is especially remarkable, given the fact that the sources for the fourth century regarding matters of Christian theology are relatively abundant and informative. Many Christian scholars since Usener's fundamental treatment have supposed that in 354 or so the pope of Rome instituted the celebration of Jesus' birth on December 25, and that as a result of his preeminent position within the Christian world, his decision was gradually accepted. Even so, the silence of our sources strikes this author as deafening. Given the multitude of doctrinal controversies that raged during these years and split Christians into numerus sects, it is surprising that our extant sources reveal so little about the change from January 6 to December 25. Why?

Could it have been that December 25 as Jesus' birthday resulted from a popular groundswell, and that the Christian elite simply made a virtue of necessity by quietly authorizing the change? If so, two possible conclusions could be drawn from such circumstances. The transition was the direct product of the widespread popularity of solar worship and its longtime overlap with the concept of the Christian deity. Moreover, our Christian sources, composed by members of the educated elite, are relatively silent on the matter, because the latter were uneasy and somewhat embarrassed about the obvious pagan associations of December 25, but in face of its popularity as Jesus' birthday in Rome and the Latin West, they felt obliged to acquiesce. In fact, Epiphanius (51.22.4–7), writing c.375 when December 25 was slowly but steadily making inroads into the Greek eastern tradition of January 6, denounced the innovation as a pagan abomination and mounted a very lengthy and spirited defense of the hallowed eastern tradition, stating (51.22.4) that Jesus was born "on the eighth day before the ides of January [January 6], thirteen days after the winter solstice and the beginning of the increase of the light and day." His wording is quite emphatic in showing that Jesus' birth came not on the winter solstice celebrated by the pagans, but thirteen days later. His defense of January 6 as the true date, however, was to no avail. By 378 or so, as we learn from a sermon of John Chrysostom (*Hom. in Nat. Christi* = *Patrologia Graeca* XLIX. 351–63), Christmas began to be celebrated in Antioch on December 25. The elaborate (and quite spurious) arguments employed by the bishop to justify the innovation suggests that many eastern Christians were hesitant to accept the new day and required a good bit of persuasion. In the end, however, the two other great cities of the Greek East, Constantinople and Alexandria, also fell into line behind Rome: Constantinople at about the same time as Antioch, and Alexandria not until sometime during the early fifth century.[75] Jerusalem came along even later. Christmas was first introduced there by Bishop Juvenal (425–458), but the celebration might not have become well established until a century or so later. Our last ancient Roman calendar, the calendar of Polemius Silvius, composed in Gaul and dating to the year 449, unlike Furius Philocalus a century earlier, unambiguously records December 25 as Jesus' birthday: "natalis Domini corporalis, solstitium, et initium hiberni = physical birthday of the Lord, solstice, and beginning of winter" (Degrassi 1963 275).[76] Yet, the dichotomy of Epiphany (January 6) first arising in the Greek East followed much later by Christmas (December 25) in the Latin West is still reflected in religious practices of modern times with the Roman Catholic Church and its descendants giving great prominence to Christmas, whereas the Greek Orthodox Church still highlights Epiphany.

Finally, the following passage, taken from a sermon of Pope Leo the Great and delivered on Christmas Day in the middle of the fifth century, offers striking testimony to the enduring influence that solar worship exercised upon Christians long after all forms of paganism had been outlawed by imperial decree:

> From such beliefs and customs arises also this impiety, namely, the adoration of the rising sun by simple folk at daybreak from an elevated place, an act which even certain Christians perform in the belief that such conduct is religious and devout. Before they arrive at the basilica of the blessed apostle Peter, which is dedicated to the one, living and true God, after they have mounted the steps which lead to the higher court, facing about they turn to the rising sun and bow head and shoulders in honor of the resplendent orb. This is done partly through ignorance, partly under the influence of Paganism, and deep is our chagrin and grief over it, because although some perhaps pay homage to the Creator of the beautiful light rather than to the light itself, a created thing, nevertheless, the very appearance of such veneration must be avoided, which, if found among us by one who has abandoned the worship of the gods, will be retained by him as a commendable part of his former beliefs, since he sees it shared by Christian and Pagan alike.[77]

EPILOGUE

In 1939 there was discovered beneath the Basilica of St. Peter in the Vatican the remains of an ancient cemetery, dating from the middle of the second century of our era to the early fourth. The discovery was made as the result of Pope Pius XI expressing the wish to be buried near Pope Pius X, who was known to have been interred beneath the church.[78] During the later Renaissance, when the modern Basilica of St. Peter was being constructed, workmen had come upon this cemetery, but until 1939 nothing was known of it other than brief mentions in the written accounts of the Renaissance. Once the modern rediscovery was made, Pope Pius XI authorized a systematic archaeological investigation, which continued for several years.

The cemetery had come into being when the race course of Gaius Caligula and Nero (Circus Gaii et Neronis = Gaianum), located in the private gardens of the Caesars, was abandoned. The cemetery itself went out of use when Constantine decided to erect over the site a church to St. Peter, because the cemetery contained a tomb that had long been regarded as housing the mortal remains of the sainted disciple. Constantine constructed his Basilica of St. Peter, so that the altar stood directly over this tomb, but in order to build the structure, workmen used rubble to cover over the cemetery to form a solid platform on which to erect the church. It was this cemetery encased in rubble that workmen in 1939 had come upon.

The excavations revealed a rich variety of funerary art, taken from both the Greco-Roman and Biblical religious traditions. Tomb M was a vaulted structure, dating to the early fourth century. Three of its interior walls were

Figure 6.1 Jesus as the Sun God (Christos Helios) From Tomb M in the ancient cemetery beneath the Basilica of St. Peter in the Vatican (Source: Public Domain Image taken from http://en.wikipedia.org/wiki/File:ChristAsSol.jpg)

painted with Biblical motifs: Jonah being swallowed by the great fish, a man standing on rocks and bringing in a fish that he has just caught, and a depiction of the good shepherd carrying a lamb over his shoulder. The vault of the tomb was decorated with a mosaic, a man driving a chariot. Since his head wears a crown with sunbeams, he is clearly identified as Christos Helios, riding across the sky.

Notes

NOTES TO CHAPTER 1

1. For a full exposition of the epigraphic texts with commentary see Degrassi 1963. For full treatments of the history, structure, and organization of the Roman calendar see Michels 1967, Radke 1990A, and Rüpke 1995. For the Roman calendar in the larger context of Greek and Roman time reckoning see Samuel 1972, Bickerman 1980, and Hannah 2005.
2. Livy I.19.6–7, Ovid *Fasti* I.27–45, III.120–127, 152–154, Plut. *Numa* 18–19, *Quaest. Rom.* 19, Gell. III.16.16, Solinus I.36, Censorinus 20.2–4 (citing M. Fulvius Nobilior, Junius Gracchanus, and Varro), Serv. *ad Georg.* I.43, and Macrob. *Sat.* I.12.3 and 13.1–3. For further treatment of the notion of a ten-month year see Rüpke 1995 192–202.
3. Nilsson 1920 54, 58, 90, 195–196, 223, and 238. Frazer 1929 II. 8–29. This explanation of Romulus' ten-month calendar was accepted by Rose 1926 90–93 and Allen 1947/8 163–164.
4. Dion. Hal. X.59.1, XVI.3.2, Plut. *Quaest. Rom.* 24, Macrob. *Sat.* I.15.5–7, and 15.15.
5. Varro *L. L.* VI.27–28, Serv. Auctus *ad Aen.* VIII.654, and Macrob. *Sat.* I.15.9–13.
6. For synthesis and analysis of these data see Forsythe 2005 82–93.
7. For the early history of the alphabet see Healey 1990 and Powell 1991 5–118. For the origin, diffusion, and other issues of the Etruscan alphabet see Cristofani 1972 and the same author's essay in Ridgway-Ridgway 1979 373–412. For a treatment of literacy in early central Italy with particular attention paid to the question of the nature of the surviving evidence see Cornell 1991.
8. Nilsson 1920 54-–87, 109–146, 176–187, 234, and 347–354.
9. For a survey of modern views see Michels 1949 and 1967 207–220.
10. Cf. Plut. *Quaest. Rom.* 25 and 102.
11. These 48 festivals are the following: Jan. 9 Ago(alia), Jan. 11 Car(mentalia), Jan. 15 Car(mentalia), Feb. 15 Lup(ercalia), Feb. 17 Qui(rinalia), Feb. 21 Fer(alia), Feb. 23 Ter(minalia), Feb. 24 Reg(ifugium), Feb. 27 Equ(irria), March 14 Equ(irria), March 17 Lib(eralia), March 19 Qui(nquatrus), March 23 Tub(ilustrium), March 24 q(uando) r(ex) c(omitiavit), Apr. 15 For(dicidia), Apr. 19 Cer(ialia), Apr. 21 Par(ilia), Apr. 23 Vin(alia), Apr. 25 (Rob(igalia), May 9 Lem(uria), May 11, Lem(uria), May 13 Lem(uria), May 21 Ago(nalia), May 23 Tub(ilustrium), May 24 q(uando) r(ex) c(omitiavit), June 9 Ves(talia), June 11 Mat(ralia), June 15 q(uando) s(tercus) d(elatum), July 5 Pop(lifugia), July 19 Luc(aria), July 21 Luc(aria), July 23 (Nep(tunalia), July 25 Fur(rinalia), Aug. 17 Por(tunalia), Aug. 19 Vin(alia), Aug. 21 Con(sualia), Aug. 23 Vol(canalia), Aug. 25 Opi(consivia), Aug. 27 Vol(turnalia), Oct.

11 Med(itrinalia), Oct. 13 Fon(tinalia), Oct. 19 Arm(ilustrium), Dec. 11 Ago(nalia), Dec. 15 Con(sualia), Dec. 17 Sat(urnalia), Dec. 19 Opa(lia), Dec. 21 Div(alia), and Dec. 23 Lar(entinalia).

12. Unconvincing in my view is Radke's explanation for the placement of the Equirria on March 14 by resorting to a non-Roman form of time reckoning, according to which the day was considered to begin at noon rather than at midnight (1993 130–131).
13. These two festivals involved horse racing in the Campus Martius. See Varro *L. L.* VI.13, Ovid *Fasti* II.858–861, III.517–522, and Festus 71L s.v. *Equirria*.
14. The term *nundinae* is derived from the Latin word for nine (*novem*), because, as was their habit, the Romans counted off this eight-day week inclusively. For more detailed treatment of the nundinal week see Rüpke 1995 453–471.
15. See Varro *De Re Rust.* II. *praef.* 1, Cic. *De Lege Agr.* II.89, Pliny XVIII.13, Macrob. *Sat.* I.16.28–35. For the establishment of market days on a Roman senator's private estate in North Africa in 138 A.D. see the *Senatus Consultum de Saltu Beguensi* = *CIL* VIII. 11,451. For a thorough treatment of the ancient evidence concerning periodic fairs and markets in Roman Imperial times see De Ligt 1989. Webster (1916 101–123) and Nilsson (1920 324–330) have collected, *inter alia*, examples of a four-day weekly cycle from West Africa, a five-day one from southern Asia, a six-day one from Aztec Mexico, a ten-day one from Inca Peru, and, of course, the seven-day one from ancient Babylonia, which was adopted by the Hebrews and eventually also replaced the eight-day nundinal cycle among the Romans.
16. Livy I.7.3–15, IX.29.9–11, Dion. Hal. I.39–44, Strabo V.3.3, Vergil *Aen.* VIII. 184–336, Ovid *Fasti* I.543–584, Propertius IV.9, Plut. *Quaest. Rom.* 60, 90, Festus 270L ss.vv. *Potitium* and *Pinarius*, the *Origo Gentis Romanae* 6–8, and *ILS* 3402–3413.
17. For the author's arguments for possible Phoenician, Greek, and Etruscan influence upon this cult, resulting from the presence of foreign traders in the Forum Boarium at a very early date, see Forsythe 2005 119–121. Cf. Rebuffat 1966, who postulates the existence of a large foreign trading community in Rome, who established and administered this cult.
18. For modern discussions of Diana in early Latium see Gordon 1932, 1934, Alföldi 1960, 1965 47ff, and Ogilvie 1965 182–183.
19. For further discussion of the possible connection between Massiliote Artemis and Roman Diana in archaic times see Ampolo 1970.
20. Degrassi 1963 458–459, Livy II.21.7, 27.5–6, Ovid *Fasti* V.669–670, Val. Max. IX.3.6, and Festus 135L s.v. *Mais Idibus*.
21. For the different ancient views on the derivation of May's name see Varro *L. L.* VI.33, Ovid *Fasti* V.1–107, Plut. *Numa* 19.3, Censorinus 22.9, and Macrob. *Sat.* I.12.19–28 with Forsythe 1994 144–150.
22. For further discussion of this event and its likely derivation from early Roman priestly records used by the historian L. Calpurnius Piso see Forsythe 1994 311–314.
23. For full treatments of the calendar's religious festivals see Fowler 1899, Scullard 1981, and König 1991.
24. Frazer's lengthy and exhaustive commentary on Ovid's *Fasti* (1929) is still a veritable mine of information for modern investigators of Roman religion. Although containing more up-to-date bibliography, Bömer's later two volume edition (1957–1958) is the work of a classicist and philologist, not of a student of Roman religion. Fauth 1978 is a valuable survey of modern scholarship on Ovid's *Fasti* as a source book on Roman religion. Within the

past decade or so Ovid's poem on the Roman calendar has received considerable modern scholarly attention, much of it, however, concentrating upon the work in the context of Augustan literature. See Herbert-Brown 1994, Fantham 1998, Herbert-Brown 2002, Green 2004, Littlewood 2006, and Robinson 2011.

25. In seeking to explain the divine name Anna Perenna Radke (1993 134–136) has suggested that Anna derives from Oscan *amma*, meaning 'mother'. Although Radke's extensive knowledge of the Italic dialects often served him well in explicating numerous aspects of archaic Roman religious terminology, his etymology in this case seems unwarranted; and although ancient authors were generally unsophisticated in their own etymological explanations, their association of Anna with the Latin word for year (*annus*) seems entirely reasonable.
26. For the ancient traditions surrounding the *salii* see Ovid *Fasti* III.259–398, Dion. Hal. II.70–71, III.32.4, Livy I.20.4, 27.7, and Plut. *Numa* 13. For the relevant archaeological evidence see Bloch 1960 134–141. For the taboo on *salii* abroad to change their place of residence during the months of March and October see Polyb. XXI.13.10–14 (Scipio Africanus in 190 B.C.) with Balsdon 1966.
27. For further discussion on this point see Loicq 1964. Radke (1993 133–134) has offered an alternative explanation for the original meaning of Mamurius Veturius, in which the first element meant 'apportioner', and Veturius was cognate with Greek *etos* (year), so that Mamurius Veturius was not the Old Mars, but the apportioner of the year, and the two groups of twelve salii and their shields represented the division of the year into twelve months.
28. Varro *L. L.* VI.33, Ovid *Fasti* IV.19–132, Plut. *Numa* 19.2, *Quaest. Rom.* 45, 86, Censorinus 22.9–11, and Macrob. *Sat.* I.12.9–15.
29. Varro L. L. VI.33, Ovid *Fasti* V.9–106, Plut. *Numa* 19.3, *Quaest. Rom.* 86, Censorinus 22.9, and Macrob. *Sat.* I.12.19–28. Words with the *etymon mai-* are quite common in Oscan (cf. the *praenomen* Mais), and Festus (121L s.v. *Maesius*) indicates that among the Oscans there was a month called Maesius.
30. Ovid *Fasti* V.147–158, Plut. *Quaest. Rom.* 20, and Macrob. *Sat.* I.12.21–27.
31. In general see Harmon 1978B 1593–1595 and Radke 1979 166–171. For the evidence from Pompeii see Boyce 1937, Orr 1978, and Giacobello 2008. For a collection of the iconographic evidence see *LIMC* 6.1 (1992) 205–212.
32. Cato *De Agri Cult.* 141, Vergil *Georg.* I.343–350, and Tibullus II.1. For the date of the Ambarvalia, which was a movable festival and was therefore not recorded in the epigraphic calendars and is also not mentioned in Ovid's *Fasti*, see the *Menologium Rusticum Colotianum* for May = *ILS* 8745 (*segetes lustrantur*).
33. The explanation is also found in Varro *L. L.* VI.33, Ovid *Fasti* I.41, V.55–78, VI.83–88, Plut. *Numa* 19.3, *Quaest. Rom.* 86, and Macrob. *Sat.* I.12.30.
34. Since the discovery and excavation in 1977 of the temple of Mater Matuta at Satricum this goddess has received considerable attention from modern scholars. See Smith in Bispham-Smith 2000 136–155. Although her epithet Matuta suggests that she was originally a divinity of ripening and maturation, she became an all-purpose goddess of fertility, childbirth, nurturing, healing, and marriage.
35. Varro *L. L.* V.153, VI.22, Livy XXVII.37.4, Plut. *Rom.* 23.3, and Paulus ex Festo 17L s.v. *Armilustrium.*
36. Varro *L. L.* VI.13, 34, Ovid *Fasti* II.19–36, Plut. *Romulus* 21.3, *Numa* 19.5, Quaest. Rom. 68, Censorinus 22.13–15, and Macrob. *Sat.* I.13.3.

37. Radke (1993 129) derives January not from the god Janus, but from *ianua*, the Latin word for door. Since the latter was clearly related to Janus, the distinction between the two possible etymologies seems hardly significant.
38. In both the prayer recorded by Cato (*De Agri Cult.* 141) for the Ambarvalia and the *devotio* formula given by Livy (VIII.9.6) Janus and Jupiter are invoked first.
39. Livy 10.47.6–7, *Per.* 11, Ovid *Fasti* I.290–294, *Metam.* XV.622–745, and Val. Max. 1.8.2.
40. The literary and epigraphic evidence for the cult of Aesculapius is quite extensive. See Edelstein 1945. For inscriptions from the site see, for example, *ILLRP* 35–39 and *ILS* 3833–3837.
41. For further discussion of the Regifugium and its symbolic significance to the early Roman yearly cycle see Magdelain 1962 and Bianchi 2011 185–192.

NOTES TO CHAPTER 2

1. Degrassi 1963 15 (*Fasti Antiates Maiores*), 189 (*Fasti Amiternini*), and 208 (*Fasti Antiates Ministrorum Domus Augustae*). In his list of the festivals from the Roman calendar Varro (*L. L.* VI.32) also includes the *dies Alliensis*, but of course, he does not give its calendrical date.
2. Besides occurring in Macrob. *Sat.* I.16.14, the same explanation is given by Varro (*L. L.* VI.29–30 and 53) and Ovid (*Fasti* I.47–48).
3. The same matter is mentioned by Festus (292L s.v. *praeciamitatores*); and the practice is likely to be as old as the Roman Republic itself, because in the famous cippus of the lapis niger (Gordon 1983 #4), commonly dated to c. 500 B.C., mentions the *rex* and his *calator.*
4. For the information set out in this paragraph the author is summarizing the much more detailed analysis of Michels 1967 31–35, 173–187, and Chart 3 at the back of the book.
5. For further discussion of this matter, which need not concern us overly much in this study see Degrassi 1963 332ff and Michels 1967 68ff.
6. There were three days that were *nefastus* until a certain religious act had been performed, after which they were *fastus*. These *dies fisi* were March 24, May 24, and June 15. The former two are marked in the calendar as QRCF = *quando rex comitiavit fas* = when the king has officiated in the Comitium, the day is *fastus*; and the latter is labeled QSDF = *quando stercus delatum fas* = when the garbage has been carried away [from the Atrium Vestae], the day is *fastus*.
7. For example, the three days of the Lemuria of May 9, 11, and 13, which were observed not by public priests, but privately in people's homes at night, were *nefasti*. Thus, there may have been some days that were neither fish nor fowl in that they did not neatly conform to the usual meanings associated with N and NP.
8. The eight *dies intercisi* are the following: January 10 between the NP days of the Agonalia and the first Carmentalia, January 14 between the NP ides and the second Carmentalia, February 16 between the NP Lupercalia and the Quirinalia, February 26 before the first Equirria, March 13 before the second Equirria, August 22 between the NP Consualia and Volcanalia, October 14 between the NP Fontinalia and the October horse, and December 12 between the NP Agonalia and the ides.
9. Detailed modern treatments of the *dies postridiani*, *dies religiosi*, *dies atri*, the *Dies Alliensis*, and the *Dies Cremerensis*, which occupy the remainder of this

chapter, are rarely encountered. For an excellent exception to this general rule see Rüpke 1995 563–575.

10. Macrob. *Sat.* I.16.21–24: "Horum causam Gellius annalium libro quinto decimo et Cassius Hemina historiarum libro secondo referunt. 21. Anoo ab urbe condita trecentesimo sexagesimo tertio a tribunis militum Verginio, Manlio, Aemilio, Postumio, collegisque eorum in senatu tractatum, quid esset propter quod totiens intra paucos annos male esset afflicta res publica. 22. Et ex praecepto patrum L. Aquinium haruspicem in senatum venire iussum religionum requirendarum gratia dixisse Q. Sulpicium tribunum militum ad Alliam adversum Gallos pugnaturum rem divinam dimicandi gratia fecisse postridie idus Quinctiles, item apud Cremeram multisque aliis temporibus et locis post sacrificium die postero celebratum male cessisse conflictum. 24. Tunc patres iussisse, ut ad collegium pontificum de his religionibus referretur; pontificesque statuisse postridie omnes kalendas, nonas, idus atros dies habendos, ut dies neque proeliares neque puri neque comitiales essent."
11. Livy V.39, Plut. *Cam.* 22, and *Quaest. Rom.* 25.
12. Cicero (*ad Att.* IX.5.2) also stresses that the early Romans enshrined the Day of the Allia, not the day on which the Gauls captured the city.
13. Varro *L. L.* VI.14: "Quinquatrus: hic dies unus ab nominis errore observatur proinde ut sint quinque dictus. Ut ab Tusculanis post diem sextum idus similiter vocatur sexatrus et post diem septimum septimatrus, sic hic, quod erat post diem quintum idus, quinquatrus."
14. Festus 304–306L s.v. *Quinquatrus*: " . . . forma autem vocabuli eius exemplo multorum populorum Italicorum enuntiata est, quod post diem quintum iduum est is dies festus, ut aput Tusculanos triatrus et sexatrus et septematrus, et Faliscos decimatrus."
15. For the loss of the -l- in alter see Warren 1901 116–117.
16. Macrob. *Sat.* I.16.24: "sed et Fabius Maximus Servilianus pontifex in libro duodecimo negat oportere atro die parentare, quia tunc quoque Ianum Iovemque praefari necesse est, quos nominari atro die non oportet."
17. On Servilianus and his writings see Peter 1914 CLXXVII–CLXXVIII and 117–118.
18. Ovid *Fasti* I.233–253, Plut. *Quaest. Rom.* 22, 41, *Origo Gentis Romanae* 1.3–3.1, and Macrob. *Sat.* I.7.19–20.
19. Wackernagel (1923–1924) argued that the term *dies ater* as a dialectal variant for *dies alter* was originally used only in reference to the day that came immediately after the ides of the month, which, of course, in very early times corresponded to the full moon. *Dies ater* therefore was associated with the waning moon, and the entire group of 36 *dies atri* eventually grew out of this notion.
20. The same categorization is found in Justinian's *Institutes* II.1.7–10 with only slight changes made due to the demise of paganism and the dominance of Christianity.
21. Festus 348–350L s.v. *religiosus*: "religiosus est non modo deorum sanctitatem magni aestimans, sed etiam officiosus adversus homines. Dies autem religiosi, quibus, nisi quod necesse est, nefas habetur facere: quales sunt sex et triginta atri qui appellantur, et Alliensis, atque ei quibus Mundus patet. ** esse ** Gallus Aelius, quod homini ita facere non liceat, ut si id faciat, contra deorum voluntatem videatur facere. Quo in genere sunt haec: in aedem Bonae Deae virum introire; adversus ** mysticiae ** legem ad populum ferre; die nefasto apud praetorem lege agere. Inter sacrum autem et sanctum et religiosum differentias bellissime refert: sacrum aedificium, consecratum deo; sanctum murum, qui sit circum oppidum; religiosum sepulcrum, ubi mortuus sepultus aut humatus sit, satis constare ait. . . ."

22. Livy VI.1.11–12 uses the phrases *de diebus religiosis* and *eadem religio esset* in describing how the Day of the Allia resulted in the pontifical classification of the after days. For further discussion of the *dies religiosi* see Degrassi 1963 360–362 and Michels 1967 63–67.
23. Varro *L. L.* VI.29: "dies postridie kalendas, nonas, idus appelati atri, quod per eos dies nihil novi inciperent."
24. Macrob. *Sat.* I.16.21: "dies autem postriduanos ad omnia maiores nostri cavendos putarunt, quos etiam atros velut infausta appellatione damnarunt."
25. Gell. V.17.2: "pontifices decreverunt nullum his diebus sacrificium recte futurum."
26. Gell. IV.9.5: "religiosi enim dies dicuntur tristi omine infames impeditique, in quibus et res divinas facere et rem quamquam novam expediri temperandum est."
27. Macrob. *Sat.* I.16.18: "unde et Varro ita scribit: 'Mundus cum patet, deorum tristium atque inferum quasi ianua patet. Propterea non modo proelium committi, verum etiam dilectum rei militaris causa habere ac militem proficisci, navem solvere, uxorem liberum quaerendorum causa ducere religiosum est.'"
28. Paul the Deacon (Paulus ex Festo 187L s.v. *Nonarum, Iduum, Kalendarum*) states that the pontiffs decreed a general ban on marriages on *dies atri*, but he wrongly regards the dividing days, not the days immediately following them, as black days. Moreover, Cicero (*ad Att.* IX.5.2) terms the *Dies Alliensis religiosus*.
29. Livy II.48–50 and Dion. Hal. IX.18–21. The legend of the Cremera regularly receives brief mention and comment in modern works concerning early Rome, usually as constituting evidence for clan-based warfare. Interest in the story has intensified in recent years as the result of the discovery in 1977 of the *Lapis Satricanus*, dating to the late sixth century B.C. and recording a dedication to Mars erected by "the *suodales* of Poplios Valesios." For more extensive treatments of the Cremera tradition, which range from outright skepticism to cautious acceptance of the basic story see Pais 1906 168–184, Ogilvie 1965 359–361, Versnel 1980, Bremmer 1982, Richard 1988, 1989, Forsythe 2005 195–200, and Smith 2006 150–153 and 290–295.
30. Livy II.49.8, Ovid *Fasti* II.201–4, Festus 358L s.v. *religioni*, 450L s.v. *Scelerata Porta*, and *Vir. Ill.* 14.3–5.
31. Since the day of the Allia has already been discussed, it has not been included in this list. For Ovid's poetic handling of the battles mentioned in Book VI of his *Fasti* see Littlewood 2006 *ad locc.*
32. Macrob. *Sat.* I.16.19–20: "Vitabant veteres ad viros vocandos etiam dies qui essent notati rebus adversis; vitabant etiam ferias, sicut Varro in augurum libris scribit in haec verba: 'Viros vocare feriis non oportet. Si vocavit, piaculum esto.' 20. Sciendum est tamen eligendi ad pugnandum diem Romanis tunc fuisse licentiam, si ipsi inferrent bellum. At cum exciperent, nullum obstitisse diem, quo minus vel salutem suam vel publicam defenderent dignitatem: quis enim observationi locus, cum eligendi facultas non supersit?"
33. Paul the Deacon (Paulus ex Festo 253L s.v. *Proeliares dies*) indicates that there were certain days on which it was not proper (*nefas*) "to provoke the enemy in war" (*hostem bello lacessere*). For a general treatment of religious scruples regarding military operations among the Greeks, Romans, and Jews see Holladay-Goodman 1986.
34. For Lucullus' Armenian campaign, which culminated in this battle see Eckhardt 1910, Ooteghem 1959 117–138, and Keaveney 1992 99–122.
35. For a list of the games established in honor of emperors and recorded in the epigraphic calendars see Degrassi 1963 373–375.
36. For the Feriale Duranum see Fink-Hoey-Snyder 1940 with Nock 1952. For the calendars of Furius Philocalus and Polemius Silvius see Degrassi 1963 237–276.

NOTES TO CHAPTER 3

1. In addition to the works discussed and cited below see H. Steuding's article on the *argei* in Roscher 1884 I. 496–500, Rose 1924 98–101, Groth 1929, Latte 1967 412–414, Holland 1961 313–331, Dumézil 1966 449, Hallet 1970 223–226, Maddoli 1971, Harmon 1978A 1447–1459, and Scullard 1981 119–121.
2. Dion. Hal. I.38, Varro *L. L.* V.45, VII.44, Ovid *Fasti* V.621–662, Plut. *Quaest. Rom.* 32, 86, Festus 450–452L s.v. *sexagenarios de ponte*, and Macrob. *Sat.* I.11.47; cf. Paulus ex Festo 18L s.v. *argea* and 66L s.v. *depontani.* Both Cicero (*Pro Rosc. Am.* 100) and Catullus (17) humorously allude to the early Romans' supposed practice of tossing elderly men from the bridge into the Tiber.
3. G. Wissowa, "Argei" in *RE* II. 1 1895 cols. 689–700.
4. For discussion of the ancient tradition surrounding these games see Ogilvie 1965 740–741, and Forsythe 1994 178–183.
5. The existence of the *collegium Capitolinorum* is well attested for the late Republic. See *ILLRP* 696–697 and Cic. *Ad Q. Frat.* II.5.2.
6. On the purgative function of the Tiber in these and other rites see Le Gall 1953 72ff.
7. J. G. Frazer has recorded customs of other peoples strikingly similar to the Roman Lemuria, including the observance of three days, the offering of beans, and the making of noise to drive away the ghosts. See Frazer 1929 IV. 41–44, which refers to additional parallel rites catalogued in his *Golden Bough.* On pp.87–89 of the former volume, he also describes a West-African rite of purification involving the setting up of effigies weeks in advance, the driving out of spirits by nocturnal noise making, and finally ending with the effigies being thrown into a nearby river.
8. For a lengthy catalogue of various peoples making offerings to rivers see Frazer 1929 IV. 97–107. Frazer in fact preferred to interpret the rites of the *argei* in these terms. He supposed that the puppets represented the various curiae of the early Roman populace.
9. On the question of the number in Varro's text see Frazer 1929 IV. 75–76.
10. For a detailed topographical discussion of these chapels see Jordan 1871 237–290.
11. This was also the view of W. Warde Fowler. See Fowler 1902B.
12. Festus 186L s.v. *novendiales*, Livy I.31.4, XXI.62.6, XXIII.31.15, XXV.7.9, XXVI.23.7, XXVII.37.1–4, XXX.38.9, XXXIV.45.8, XXXV.9.5, XXXVI.37.5, XXXVIII.36.4, XXXIX.22.3, and XLIV.18.6.
13. Livy XXVII.37 (207 B.C.), XXXI.12.6–10 (200 B.C.), Julius Obsequens 27A (133 B.C.), 34 (120 B.C.), 36 (117 B.C.), 46 (99 B.C.), 48 (97 B.C.), and 53 (92 B.C.).
14. For evidence of this belief among the Romans see Livy III.58.11 (the murdered Verginia), Suet. *Otho* 7.2 (the assassinated Emperor Galba), and Pliny *Ep.* VII.27 (a collection of three stories). Cf. Plut. *Caes.* 69.5–14 concerning a phantom that appeared to Brutus, one of Caesar's assassins. On the malevolent character of the Lemuria see Rose 1944.
15. E.G., adjectives: adoreus, aeneus, aequoreus, arboreus, argenteus, aureus, caeruleus, corneus, cupreus, eburneus, farreus, ferreus, frondeus, hordeus, idoneus, igneus, iunceus, lapideus, ligneus, linteus, niveus, piceus, querceus, roboreus, sanguineus, squameus, taureus, vitreus.

    E.G., nouns: alveus, balteus, clipeus, culeus, maleus, pileus, pluteus, puteus, urceus.
16. Varro *L. L.* VII.44, Ovid *Fasti* V.621–622, 659, and Festus 450–452L.
17. Cato *De Agri Cult.* 10.3, 11.4, Varro *De Re Rust.* I.23.5, Pliny *NH* VII.206, and XVI.178.
18. On the etymology of *argentum* and *argilla* see Walde-Hofmann 1938 I. 66.

19. Mannhardt 1875 311ff. Tacitus in *Germania* 40.5 records a Germanic custom involving the immersion into a lake of a goddess named Nerthus and the drowning of her ministrants in the same body of water.
20. See, for example, Taylor 1966 92.
21. Fowler 1902 119, 1899 120, and Rose 1924 99.

NOTES TO CHAPTER 4

1. The *Ludi Saeculares* receive general treatment in most modern works on Roman religion, but their peculiar nature and obscure origin have attracted much modern scholarly study and speculation, on which see Roth 1853, Fowler 1911 438–447, M. P. Nilsson's article "Saeculares Ludi" in *RE* I.2 1920 1696–1719, Frazer 1929 II. 191–203, Wuilleumier 1932, Taylor 1934, Wuilleumier 1938, Wagenvoort 1956 193–232, Latte 1967 246–248, Pighi 1965, Palmer 1974 94–108, Brind'Amour 1978, Pascal 1979, Moretti 1985, Coarelli 1993, Forsythe 1994 399–404 with 489–494, Bernstein 1998 129–142, Feeney 1998 127–133, and Orlin 2010 67–71. The author has been unable to obtain Schnegg-Köhler 2002.
2. In order to maintain the link between the Roman concept of the *saeculum* and the *Ludi Saeculares*, the author has decided to adopt the peculiar spelling 'saecular' instead of the misleading 'secular' in the English phrase Saecular Games'.
3. Val. Max. II.4.5: "Et quia ceteri ludi ipsis appellationibus unde trahantur apparet, non absurdum videtur saecularibus initium suum, cuius generis minus trita notitia est, reddere. Cum ingenti pestilentia urbs agrique vastarentur, Valesius, vir locuples rusticae vitae, duobus filiis et filia ad desperationem usque medicorum laborantibus, aquam calidam iis a foco petens, genibus nixus Lares familiares, ut puerorum periculum in ipsius caput transferrent, oravit. Orta deinde vox est, habiturum eos salvos, si continuo flumine Tiberi devectos Tarentum portasset, ibique ex Ditis Patris et Proserpinae ara petita aqua[A] recreasset. Eo praedicto magnopere confusus, quod et longa et periculosa navigatio imperabatur, spe tamen dubia praesentem metum vincente, pueros ad ripam Tiberis protinus detulit: habitabat enim in villa sua propter vicum Sabinae regionis Eretum. Ac lintre Ostiam petens, nocte concubia ad Martium Campum appulit. Sitientibusque aegris succurrere cupiens, igne in navigio non suppetente, ex gubernatore cognovit haud procul apparere fumum; et ab eo iussus egredi Tarentum[B] (id nomen ei loco est), cupide adrepto calice aquam flumine haustam eo, unde fumus erat obortus, iam laetior pertulit, divinitus dati remedii quasi vestigia quaedam in propinquo nanctum se existimans, inque solo magis fumante quam ullas ignis habente reliquias, dum tenacius omen adprehendit, contractis levibus et quae fors obtulerat nutrimentis, pertinaci spiritu flammam evocavit, calefactamque aquam pueris bibendam dedit. Qua potata, salutari quiete sopiti diutina vi morbi repente sunt liberati. Patrique indicaverunt vidisse se in somnis a nescio quo deorum spongea corpora sua pertergeri et praecipi, ut ad Ditis Patris et Proserpinae aram, a qua potio ipsis fuerat adlata, furvae hostiae immolarentur lectisterniaque ac ludi nocturni fierent. Is, quod eo loci mullam aram viderat, desiderari credens, ut a se constitueretur, aram empturus in urbem perrexit, relictis qui fundamentorum constituendorum gratia terram ad solidum foderent. Hi domini imperium exequentes, cum ad XX pedum altitudinem humo egesta pervenissent, animadverterunt aram Diti Patri Proserpinaeque inscriptam. Hoc postquam Valesius nuntiante servo accepit, omisso emendae arae proposito, hostias nigras, quae antiquitus furvae dicebantur, Tarenti[C] immolavit

ludosque et lectisternia continuis tribus noctibus, quia totidem filii periculo liberati erant, fecit. Cuius exemplum Valerius Publicola, qui primus consul fuit, studio succurrendi civibus secutus apud eandem aram, publice nuncupatis votis caesisque atris bubus, Diti maribus, feminis Proserpinae, lectisternioque ac ludis trinoctio factis, aram terra, ut ante fuerat, obruit."

A. Some manuscripts read *calda* instead of *aqua*, apparently in order to stress the need for the curative water to be hot and thus to require the use of fire.

B. Some manuscripts read *Terentum* instead of *Tarentum*, thereby suggesting that the toponym derived from *terra*.

C. Some manuscripts read *Terenti* instead of *Tarenti* for the reason just stated.

4. Since Valerius Maximus in the passage quoted above styles Publicola the first consul, we might suppose that the author intended to date Publicola's performance of the rites at the Tarentum to his first consulship in 509 B.C. If so, this agrees with Zosimus (II.3). The text of Censorinus 17.10, which cites Valerius Antias for the first celebration of the *Ludi Saeculares* by Publicola, is lacunose, so that there is possible doubt as to when Censorinus wished to date this event. Wiseman (1998 167) plausibly argues that Censorinus' wording, despite the lacuna, can be interpreted to mean that Censorinus (and hence Valerius Antias) dated Publicola's observance of the rites to the first year of the Republic.
5. Taylor (1934) also regards Valerius Antias along with members of the Valerian family as having been very influential in shaping the historical traditions surrounding the Saecular Games.
6. Censorinus' text at 17.10 is lacunose where he is contrasting the variant dates of Valerius Antias vs. the *Commentarii* of the Augustan *quindecimviri sacris faciundis* for the supposed second observance of the Saecular Games. Nevertheless, the overall context of the whole passage makes it clear that Antias and the priestly records differed from one another by two years; and since Antias dated the games according to a 100-year cycle, it is virtually certain that his date in this section of Censorinus was 348 B.C., the first consulship of the great M. Valerius Corvus, whereas the *quindecimviri sacris faciundis* of Augustus' day dated them two years later (346 B.C.) to Corvus' second consulship. The lacuna in the text doubtless resulted from haplography caused by the double occurrence of Corvus' name.
7. Varro *L. L.* VI.11: "seclum spatium annorum centum vocarunt, dictum a sene, quod longissimum spatium senescendorum hominum id putarunt."
8. Although Claudius' celebration of the *Ludi Saeculares* seems to have been based upon a 100-year *saeculum*, Hirschfeld (1881 102) observed that the Claudian games occurred 550 years after the supposed arrival of the Claudian family in Rome from the Sabine territory, which therefore would have been five *saecula* of 110 years before Claudius observed the Saecular Games.
9. See Lanciani 1897 446–448. For the exact location of the Tarentum see Map 14 of Lanciani's *Forma Urbis Romae*.
10. Palmer (1990 Plan 2 between pages 13 and 14) locates the Tarentum in the area to the west of the Ponte Sant'Angelo, whereas Coarelli (1993 234 Fig. II) places it just to the south of the Ponte Vittorio Emanuele.
11. Despite other modern scholars' preference for the form Tarentum vs. Terentum attested in ancient sources, the author sides with Wagenvoort (1956 198–204) in regarding Terentum as the more likely original spelling of the toponym, and that it simply had the obvious meaning of the place where an altar was covered by earth.
12. For detailed discussion of the traditions surrounding the Lacus Curtius see Bremmer 1993 165–170, Forsythe 1994 149–157, and Oakley 1998 96–102.

13. In explaining the term *ad Murciae* for the inner part of the Circus Maximus Varro (*L. L.* V.154) uses the phrase "ut Procilius aiebat = as Procilius used to say." This suggests that Procilius was an older contemporary and acquaintance of Varro, and that by the 40s B.C. he was dead when Varro was writing his *De Lingua Latina*.
14. See Plut. *Numa* 1.2 with Livy VI.1.1–3 and Oakley 1997 381–382.
15. See F's 7, 10, 12, 15, 42–43, 56, 57A, 64, 69, and 80. For Livy's use of Quadrigarius in his second pentad see Forsythe 2007.
16. Since this matter has a direct bearing upon the early chronology of Roman *saecula*, throughout the remainder of this chapter the author will designate this year as 362/358 B.C. so as to indicate both its Varronian and absolute date.
17. Censorinus 17.13: "Roma condita anno DC septimum saeculum occipit his consulibus, qui proximi sunt consules, M. Aemilius M. f. Lepidus, C. Popillius II absens."
18. Serv. Auct. *ad Aen.* II.140: "Si gravida [hostia] fuerat, forda dicitur. Quae sterilis autem est, taurica appellatur, unde Ludi Taurei dicti, qui ex Libris Fatalibus a rege Tarquinio Superbo instituti sunt, propterea quod omnis partus mulierum male cedebat. Alii Ludos Taureos a Sabinis propter pestilentiam institutos dicunt, ut lues publica in has hostias verteretur." Varro (*De Re Rust.* II.5.6) confirms that the proper term for an infertile cow was *taura*.
19. Varro *L. L.* V.143, *De Re Rust.* II.1.10, Serv. *ad Aen.* IV.212, V.755 (quoting Cato's *Origines*), Dion. Hal. I.88.2, Ovid *Fasti* IV.825–826, Columella VI. *Praef.* 7, Plut. *Rom.* 11, *Quaest. Rom.* 27, Festus 270–271L s.v. *primigenius sulcus*., and Zonaras VII.3. For modern discussion of the importance of the *pomerium* and the plowing of the *sulcus primigenius* in establishing colonies as small replicas of Rome see Gargola 1995 74–81.
20. Varro L. L. V.154: "Item simili de causa Circus Flaminius dicitur, qui circum aedificatus est Flaminium Campum, et quod ibi quoque Ludis Tauriis equi circum metas currunt."
21. Pseudo-Acro in Hor. *Carmen Saec.* 8: "Valerius [Verrius] Flaccus refert Carmen Saeculare et sacrificium inter annos centum et decem Diti et Proserpinae constitutum bello Punico primo ex responso decemvirorum, cum iussi essent Libros Sibyllinos inspicere ob prodigium, quod eo bello accidit: nam pars murorum urbis fulmine icta ruit. Atque ita responderunt bellum adversus Carthaginienses prospere geri posse, si Diti et Proserpinae triduo (id est, tribus diebus et tribus noctibus) ludi fuissent celebrati et carmen cantatum inter sacrificia. Hoc autem accidit consulibus P. Claudio Pulchro, L. Iunio Pullo [249 B.C.], cum Roma pestilentia laboraret. Ex Libris Sibyllinis iussum est, ut Diti Patri ad Terentum stipes mitteretur. Hoc etiam idem libri iusserunt, ut nobilium liberi in Capitolio hoc carmen decantarent."
22. Censorinus 17.8: "Varro *De Scaenicis Originibus* libro primo ita scriptum reliquit: 'Cum multa portenta fierent, et murus ac turris, quae sunt intra Portam Collinam et Esquilinam, de caelo tacta essent, et ideo Libros Sibyllinos decemviri adissent, renuntiarunt, uti Diti patri et Proserpinae Ludi Tarentini in Campo Martio fierent tribus noctibus, et hostiae furvae immolarentur, utique ludi centesimo quoque anno fierent.'"
23. Our most important ancient source for the First Punic War is, of course, Book I of Polybius, an accurate summary of the contemporary and nearly contemporary accounts of Philinus of Acragas and Q. Fabius Pictor. For a detailed, recent modern treatment of the war see Lazenby 1996.
24. For such an example see the chapter quoted from Livy at the beginning of the next section of this chapter.
25. Livy *Per.* XIX, Val. Max. VIII.1. *Damnati* 4, Suet. *Tib.* 2.3, and Gell X.6. For a detailed discussion of this curious incident see Suolahti 1977.

26. For collections made to honor public figures see Livy II.33.11, III.18.11, Val. Max. IV.4.2, and Pliny XXXIII.138. For *stipes* as expressions of popular piety see Livy V.25.5, XXII.1.18, XXV.12.14, Cic. *De Leg.* II.22, and 40. It must have been common for people to make modest offerings to divinities in sacred groves, because one explanation for the derivation of *lucrum* (= profit) was that it originally referred to the money collected from *stipes* accumulated in sacred groves (= Latin *luci*). See Paulus ex Festo 106L s.v. *lucaris pecunia* and Plut. *Quaest. Rom.* 88.
27. Suet. *Aug.* 57.1; cf. *Caligula* 42. *ILS* 92 and 93 come from inscribed statue bases and show that Augustus used money collected by the people as a *stips* to honor him on New Year's Day to erect statues for Mercury and Volcan.
28. For further treatment of this series of religious ceremonies see Boyce 1937 and MacBain 1982 65–71. For the Roman expiation of deformed human offspring see Allély 2003.
29. Livy XXVII.37: "1. Priusquam consules proficiscerentur, novendiale sacrum fuit, quia Veiis de caelo lapidaverat. 2. Sub unius prodigii, ut fit, mentionem alia quoque nuntiata: Minturnis aedem Iovis et lucum Maricae, item Atellae murum et portam de caelo tacta; 3. Minturnenses, terribilius quod esset, adiciebant sanguinis rivum in porta fluxisse; et Capuae lupus nocte portam ingressus vigilem laniaverat. 4. Haec procurata hostiis maioribus prodigia, et supplicatio diem unum fuit ex decreto pontificum. Inde iterum novendiale instauratum, quod in Armilustro lapidibus visum pluere. 5. Liberatas religione mentes turbavit rursus nuntiatum Frusinone natum esse infantem quadrimo parem; nec magnitudine tam mirandum quam quod is quoque, ut Sinuessae biennio ante, incertus mas an femina esset natus erat. 6. Id vero haruspices ex Etruria acciti foedum ac turpe prodigium dicere: extorrem agro Romano, procul terrae contactu, alto mergendum. Vivum in arcam condidere provectumque in mare proiecerunt. 7. Decrevere item pontifices, ut virgines ter novenae per urbem euntes carmen canerent. Id cum in Iovis Statoris aede discerent conditum ab Livio poeta carmen, tacta de caelo aedis in Aventino Iunonis Reginae; 8. prodigiumque id ad matronas pertinere haruspices cum respondissent donoque divam placandam esse, 9. aedilium curulium edicto in Capitolium convocatae quibus in urbe Romana intraque decimum lapidem ab urbe domicilia essent. Ipsae inter se quinque et viginti delegerunt, ad quas ex dotibus stipem conferrent. 10. Inde donum pelvis aurea facta lataque in Aventinum, pureque et caste a matronis sacrificatum. 11. Confestim ad aliud sacrificium eidem divae ab decemviris edicta dies, cuius ordo talis fuit. Ab aede Apollinis boves feminae albae duae porta Carmentali in urbem ductae. 12. Post eas duo signa cupressea Iunonis Reginae portabantur; 13. tum septem et viginti virgines, longam indutae vestem, carmen in Iunonem Reginam canentes ibant, illa tempestate forsitan laudabile rudibus ingeniis, nunc abhorrens et inconditum si referatur. Virginum ordinem sequebantur decemviri coronati laurea praetextatique. 14. A porta Iugario Vico in Forum venere. In Foro pompa constitit; et per manus reste data, virgines sonum vocis pulsu pedum modulantes incesserunt. 15. Inde Vico Tusco Velabroque per Bovarium Forum in Clivum Publicium atque aedem Iunonis Reginae perrectum. Ibi duae hostiae ab decemviris immolatae, et simulacra cupressea in aedem inlata."
30. Livy XXXI.12.6–10): "6. Iam animalium obsceni fetus pluribus locis nuntiabantur. In Sabinis incertus infans natus, masculus an femina esset, alter sedecim iam annorum item ambiguo sexu inventus. 7. Frusinone agnus cum suillo capite, Sinuessae porcus cum capite humano natus, in Lucanis in agro publico eculeus cum quinque pedibus. 8. Foeda omnia et deformia errantisque in alienos fetus naturae visa. Ante omnia abominati semimares iussique in mare extemplo deportari, sicut proxime C. Claudio M. Livio consulibus deportatus

similis prodigii fetus erat. 9. Nihilo minus decemviros adire libros de portento eo iusserunt. Decemviri ex libris res divinas easdem, quae proxime secundum id prodigium factae essent, imperaverunt. Carmen praeterea ab ter novenis virginibus cani per urbem iusserunt donumque Iunoni Reginae ferri. 10. Ea uti fierent, C. Aurelius consul ex decemvirorum responso curavit. Carmen, sicut patrum memoria Livius, ita tum condidit P. Licinius Tegula."

31. Appian's narrative in *Punica* 127ff indicates that Carthage was finally captured and destroyed in the spring of 146. Thus, unless the Roman calendar was drastically out of phase with the natural yearly cycle, the news is likely to have reached Rome before the celebration of the *Ludi Saeculares*.
32. Serv. and Serv. Auct. *ad Ecl.* IX.46: "Cum Augustus Caesar ludos funebres patri celebraret, die medio stella apparuit. ille eam esse confirmavit parentis sui. . . . *Baebius Macer circa horam octavam stellam amplissimam, quasi lemniscis, radiis coronatam, ortam dicit. quam quidam ad inlustrandam gloriam Caesaris iuvenis pertinere existimabant, ipse animam patris sui esse voluit eique in Capitolio statuam, super caput auream stellam habentem, posuit: inscriptum in basi fuit 'Caesari emitheo'. sed Vulcanius aruspex in contione dixit cometen esse, qui significaret exitum noni saeculi et ingressum decimi; sed quod invitis diis secreta rerum pronuntiaret, statim se esse moriturum: et nondum finita oratione, in ipsa contione concidit. hoc etiam Augustus in libro secundo de memoria vitae suae complexus est. . . .*" For further discussion of saecular omens at this time see Weinstock 1971 191–196.
33. On this very complex theme and its possible influence on the Roman *saeculum* and *Ludi Saeculares* see Alföldi 1977 and Brind'Amour 1978.
34. The inscribed remains of these chapels from Rome are quite plentiful. See, for example, *ILS* 3612–3622 with Taylor 1931 184–189 and Lott 2004 81–220.
35. For these texts and their explication see the progressively more complete editions of Henzen 1874, Pasoli 1950, and Scheid 1990. See also the discussions of Olshausen 1978, Beard 1985, and Linderski 1995 600–602. In s.7 of the *Res Gestae* Augustus lists his membership of this priesthood among other religious offices that he held.
36. In this paragraph the author summarizes the clear and convincing work of A. H. M. Jones in his 1951 article concerning the evolution of Octavian/Augustus' legal position within the Roman state.
37. See Dio LIV.12–14 for Agrippa's designation as partner and successor and for the *lectio senatus*. For a full list of ancient references to these important laws see Rotondi 1966 443–447.
38. This paragraph confines itself to summarizing the rites described in *ILS* 5050 for the three days of the Saecular Games proper. It excludes from treatment the gods listed in the new epigraphic material adduced by Moretti (1985 370). The latter contains what appears to be a fragmentary text of an official letter addressed to the senate from Augustus, acting as *magister* on behalf of the *quindecimviri*, which describes the performance of a prayer to several divinities, who, according to our previous state of knowledge, were not associated with the *Ludi Saeculares*. Moretti dates this letter between February 17 and March 24 of 17 B.C, coming between two other dated senatorial decrees. Thus, the ceremony recorded in this letter seems to have been a ritual that formally announced the season of the *Ludi Saeculares* and, as the list of divinities indicates, involved the invocation of several gods and goddesses who did not figure directly in the rites of the three-day period.
39. For the coin of Augustus see Mattingly and Sydenham 1923 75 with pl. I. 16. For the Domitianic coins commemorating his Saecular Games see Mattingly and Sydenham 1926 201–202 with no. 376 portraying the distribution of *suffimenta*.

NOTES TO CHAPTER 5

1. Modern bibliography on the Phrygian cult of Cybele and Attis, the corresponding Roman cult of Magna Mater, and the taurobolium is considerable. The standard older works are Showerman 1901, Hepding 1903, and Graillot 1912. Two more recent, excellent surveys are Vermaseren 1977 and Thomas 1984. Lane 1994 is a collection of studies written by different scholars on assorted topics. Roller 1999 is another excellent study that primarily concentrates upon the archaeological and art-historical evidence for the cult from the archaic period to the time of Augustus. Attis is treated in the larger context of dying and reviving gods in volume IV of Sir James Frazer's classic work, *The Golden Bough*, London 1907. More general studies can be found in surveys of ancient religions during Hellenistic and Roman Imperial times. Note especially Chapter 3 of Cumont 1929 and Chapter 1 of Turcan 1996. The iconographic evidence for the cult is treated by Vermaseren 1966 and more exhaustively in idem 1977–1989. Two revisionist studies of the taurobolium are Rutter 1968 and Duthoy 1969. A minimalist approach to the cult of Cybele and Attis in the Roman world has also been adopted by Lambrechts 1962 and 1967.
2. Lactantius (*De Mortibus Persecutorum* 12.1) states that Diocletian deliberately chose February 23, the Terminalia, as the day on which to begin the persecution because it would be auspicious in exterminating the religion of the Christians. For further details concerning persecution of Christians and, Constantine's conversion, and the conflict between Christianity and paganism during the fourth century see Chapter 6.
3. When used in lowercase, 'imperial' will signify 'belonging or pertaining to the emperor', whereas the capitalized form of the word will refer to the Roman Empire. Although this may seem trivial, the distinction is an important one, especially when treating the taurobolium.
4. For the text of this imperial pronouncement see Lactantius ibid. 48 and Eusebius *HE* X.5.2–14.
5. For the phenomenon of the mystery religions during Hellenistic and Roman Imperial times see in general Cumont 1929, Reitzenstein 1978, Martin 1987, Burkert 1987, Turcan 1996, and Bowden 2010. *ANRW* II. 17.3–4 contain major articles with extensive bibliographies on the mystery religions of the Roman Empire.
6. CIL VI. 497–504, dated respectively as follows: April 4, 305; April 29, 350; July 19, 374; May 13, 377; April 5, 383; April 5, 383; May 23, 390; and August 13, 376.
7. For a modern discussion of these taurobolic inscriptions in the broader political and religious context of the late fourth century A.D. see Matthews 1973.
8. The texts of most of these inscriptions are to be found scattered throughout the various volumes of *CIL*, and others have been published in different journals and epigraphic publications, but a complete collection of this material is now available more conveniently in Duthoy 1969 7–53.
9. These inscriptions are: *IGRRP* IV. 294 = *OGIS* 764 (Pergamum, ca.135 B.C.); *AJA* 39 (1935) pp.589–591 (Ilium, first century B.C.); *TAM* II. 508 (Pinara in Lycia, first century B.C.); *IGRRP* IV. 494+499–500 (Pergamum, ca.105 A.D.); and *CIL* X. 1596 (Puteoli, 134 A.D.).
10. *CIL* VI. 1675, 1778, 1780, 31,940, and IX. 6099.
11. A. D. Trendall and A. Cambitoglou, *The Red-Figured Vases of Apulia*, I. (1978) no. 4/229. For the Greek rites to cleanse a person from the pollution of homicide see Frazer 1929 II. 287–289 and Parker 1983 370–374.

12. Hdt. I.34–45. A similar but much briefer account of this tale is recorded by Pausanias (VII.17.9–10), who cites as his source Hermesianax of Colophon, a writer of the early third century B.C.
13. Ovid (*Metam.* X.104) portrays Attis as transformed into a pine tree. Conversely, Plutarch (*De Iside et Osiride* 69) states that among the Phrygians the god was believed to be asleep during the winter.
14. On this point see Seyrig 1944.
15. For this phenomenon see Gaster 1961.
16. This would, of course, help to explain the prominence of the testicles of the bull or ram mentioned in some of the taurobolic and criobolic inscriptions. Cf. Clement of Alexandria, *Protrep.* II.15, where Zeus' offering to Rhea of a ram's testicles in substitution of his own forms a mythical etiology for the rites of Magna Mater. The procreative organ of a sacrificed animal is used to secure the initiate's regeneration after death.
17. For 'Adrasteia' as an epithet of Nemesis see Aeschylus *Prometheus Vinctus* 936.
18. Herodotus (IV.76) indicates that there was an important cult of Cybele in Cyzicus, near which lay Mount Dindymon, whence Cybele received the epithet 'Dindymenean'. Cf. Vermaseren *CCCA* I. 90–98. Moreover, the Plain of Adrasteia, located in Mysia on the Hellespont and through which the Granicus River flowed, took its name from a legendary figure named Adrastus, who according to Callisthenes established there an altar to Nemesis (Strabo XIII.1.11–13). Strabo (ibid.) indicates that Cyzicus had a cult to Nemesis Adrasteia.
19. Paus. VII.17.10–12 and Arnobius V.5–7. 'Adrastus' may have been subsequently employed as an epithet of the Great Mother: for an inscription discovered in the Meander Valley (*BCH* 2 [1887] 349 no.5) describes a person as "priest of the Goddess Mother Adrastus," which is perhaps to be understood as "the inexorable Mother Goddess."
20. Attis' association with swine is further suggested by the mystic exclamation "hyes Attes" mentioned by Demosthenes (*De Corona* 260) in connection with the cult of the Phrygian god Sabazius.
21. Actually, her official title found in inscriptions of the Roman state was "Mater Deum Magna Idaea, (Great Idaean Mother of the gods)" often simply abbreviated "M.D.M.I."
22. The two principal ancient accounts of this event are Livy XXIX.10.4–11.8, 14.1–14, and Ovid *Fasti* IV.247–348. For modern discussions of this affair see Köves 1963, Bömer 1964, Gruen 1990 3–30, and Orlin 2010 77–83. For evidence of diplomatic relations between Pergamum and Pessinus during the mid-second century B.C. see Welles 1934 241–253.
23. Livy XXXVI.36.3–4, *Fasti Praenestini* = Degrassi 1963 128–129, and Hadzsits 1930.
24. For Mount Ida in Crete see Diod. V.70, Strabo X.4.8, Paus. V.7.6, and Diog. Laert. VIII.1.3 (cf. Hesiod *Theogony* 477–484). For Mount Ida in the Troad see Homer *Iliad* VIII.47–48, Strabo X.7.3, and Dion. Hal. I.61.4. According to Homer (*Iliad* II.820–821) Aeneas was born and raised on Mount Ida. Vergil (*Aen.* III.104–191) incorporated the traditions of both mountains into his poetic account of Aeneas. It is worth noting that just as the infant Zeus was sustained by Amalthea's goat's milk on Mount Ida in Crete, so Attis was tended by a goat when exposed as an infant (Paus. VII.17.11 and Arnobius V.6). An archaic bronze tympanum and shield, apparently related to the worship of the Great Mother Goddess and her son Zeus, have been found in the Idaean Cave in Crete. See Dunbabin 1957 40–41 and Boardman 1980 58–60.
25. *CIL* VI. 505, a taurobolic altar dated February 26, 295 A.D.

26. Livy V.22.3–7, IV.25.3+29.7, V.13.4–8, VII.2, X.47.6–7, *Per.* 11, Ovid *Fasti* I.290–294, and *Metam.* XV.626–744.
27. See Cic. *Pro Balbo* 55 for the Greek priestess of Ceres, Dion. Hal. II.19.4 for the Phrygian priest and priestess of Magna Mater in Rome, Lucretius II.600–643 for the public processions through the city, Cic. *De Leg.* II.22 and 40 for the cult's sacred mendicancy allowed on certain days, Gell. II.24.2 and XVIII.2.11 for banquets of the Roman elite held in honor of the goddess during the Megalensian Games of April. Cf. Ovid *Fasti* IV.179–220, 349–372, Diod. XXXVI.13, Catullus 63, and Apuleius *Metam.* VIII.25–29.
28. Livy XXXIX.8–19 with *ILS* 18 = *ILLRP* 511 = Gordon 1983 #8. For modern treatments of this much discussed episode in Roman religious history see, for example, North 1979, Forsythe 1994 385–396, and Orlin 2010 165–168.
29. The demolition of altars and shrines to the Egyptian deities in Rome are recorded for 58 (Tertullian *Apologia* 6.8 and *Ad Nationes* I.10.17–18), 53 (Dio XL.47.3), and 48 B.C. (Dio XII.26.2). In 28 B.C. Octavian forbade the celebration of the Egyptian religion inside the pomerium (Dio LIII.2.4); and while Augustus was absent from Rome in 21 B.C., Agrippa extended the prohibition to an area within one mile of the city (Dio LIV.6.6). In 19 A.D., as the result of the involvement of two priests of Isis in the seduction of a Roman noble woman named Paulina by the equestrian Decius Mundus, the Emperor Tiberius ordered that the priests be crucified, their temple demolished, and the cult statue thrown into the Tiber (Tac. *Ann.* II.85, Suet. *Tib.* 36, and Josephus *Ant. Iud.* XVIII.3.4).
30. In confirmation of this last statement one should note Julius Obsequens 44A and Val. Max. VII.7.6. The former records as a prodigy for the year 101 B.C. a slave of the consular Q. Servilius Caepio castrating himself amid the rites of Magna Mater. The slave was banished abroad and not allowed to return to Rome. His treatment resembles the official Roman religious practice of ejecting hermaphroditic children from Roman territory by enclosing them in a chest and casting them into the sea. See Livy XXVII.37.5–6 and XXXIX.22.5. Valerius Maximus relates that in 77 B.C. the consul overruled a decision of the urban praetor in a case in which a Gallus of Magna Mater, named Genucius, had been instituted heir by a freedman. When the praetor granted Genucius possession of the estate in accordance with the will, the patron of the testator appealed to the consul, who subverted the testamentary disposition on the grounds that since Genucius had voluntarily deprived himself of his manhood, he should be regarded as being neither man nor woman. The conclusion to be drawn from this incident might be that in the view of the consul, a Roman citizen's self-neutering while participating in such un-Roman religious rites deprived him of his civic status.
31. The evidence for the foundation of this shrine is unclear. See Wissowa 1912 353–354 and Richardson 1992 211. For the orientalizing tendencies in Gaius' reign see Colin 1954.
32. Tac. *Hist.* IV.81–82, Suet. *Vesp.* 7, Josephus *Bell. Iud.* VII.5.4. Cf. Suet. *Dom.* 1.2 and Eutropius VII.23.5.
33. *AJA* 37 (1933) 215–270: description in Italian by A. Vogliano, commentary in French by F. Cumont, and English summary by C. Alexander.
34. An inscription from Lactora in Aquitania (*ILS* 4127), dating to March 24 of 239 A.D., records the castration of a man named Eutyches (possibly a slave): S(acrum) M(atri) D(eum) | Val(eria) Gemina | vires esce- | pit Eutyche- | tis VIIII kal. | April. sacer- | dote Traia- | nio Nundi- | nio, d. n. Gordi- | ano et Aviola cos.
35. See, for example, Fishwick 1966 193–202. Cf. Thomas 1984 1518–1520. For a detailed survey of the ancient evidence, including a critique of Lambrecht's

minimalist approach see Vermaseren 1977 110–124. For a strenuous argument dissociating Claudius altogether from the Phrygian cult see Scramuzza 1940 152–155, whose notes cite earlier scholarship.

36. For a brief but excellent discussion of Lydus' *De Mensibus* see Maas 1992 53–66.
37. Vermaseren 1966 54–55, 1977 55, and *CCCA* III. 1977 97–98 with plates CCII–CCIV. For treatments of this structure and its numerous mythological scenes see Bagnani 1919, Strong and Joliffe 1924, and Carcopino 1926.
38. See Plut. *Sulla* 9.7, 27.12, [Caesar] *Bell. Alex.* 66, Vergil *Aen.* VIII.703, and Tibullus I.6.45–50. The inscriptions from Beneventum (*CIL* IX. 1538–1542) reveal Magna Mater's warlike character by entitling her "Berecynthian Minerva." Strabo (XII.2.3+5) says that the priest of Ma in Comana was of the royal Cappadocian line and was second in importance only to the king. The temple commanded considerable financial resources, including land and thousands of sacred slaves. The Greeks regarded the cult as that of Artemis Tauropolus, imported from the Tauric Chersonnesus by Orestes and Iphigenia. This explanation of the cult's origin suggests that the Cappadocian worship of Ma included some kind of taurobolic rites. In fact, many of the taurobolic inscriptions spell taurobolium with a p instead of a b.
39. *CIL* IX. 1538 and 1541–1542 = *ILS* 4184–4185 from Beneventum. Since *CIL* VI. 502 (= *ILS* 4150) and 508 (= *ILS* 4146) record the titles "sacerdus (sic) maxima" and "sacerdotem Phrygem maximum," the cult of Magna Mater in Rome could have been served by a board or hierarchy of priests.
40. There are 36 taurobolic inscriptions that can be clearly identified as being of imperial type. The earliest one (*CIL* XIII. 1751 = *ILS* 4131, quoted and discussed below) dates to the year 160 A.D., the end of Antoninus Pius' reign, and the last datable imperial taurobolic inscription belongs to the joint reign of Diocletian and Maximian near the close of the third century (*CIL* VIII. 23,401 = *ILS* 4142, from Mactar). It is interesting to note that during the rather unsettled times of the mid-third century the formula "pro salute" was sometimes expanded to "pro salute et reditu et victoria." See *CIL* XIV. 42–43, 4303, and 4306 (all from Ostia). Three inscriptions (*CIL* II. 5521, and XIII. 1568–1569) record taurobolia performed "pro salute imperii" without specifying the emperor's name, whereas another text (*CIL* XIII. 1753 = *ILS* 4133), dating to May of 194 A.D., records a taurobolium performed for the welfare of Septimius Severus and Clodius Albinus.
41. For a brief but representative synopsis of this evidence see Vermaseren 1977 126–144. For its exhaustive documentation see the same author's *CCCA* (above, n.1).
42. *AE* 1910 #217 and 1924 #26.
43. For the various offices see Thomas 1984 1528–1533.
44. CIL X. 4726 (Forum Popilii), IX. 1538 (Beneventum), 1541 (Beneventum), and XII. 1567 (Arausio).
45. The *cognomina* of these three men were certainly regarded as auspicious. The names Pannychus and Votivus have obvious religious associations: the former being connected with the all-night vigil of the Eleusinian and Orphic Mysteries (Aristoph. *Frogs* 371, 448, Hdt. II.62, and IV.76), and the latter pertaining to vows commonly made to the gods. The name Felicianus must have been viewed as portending the felicity of the new priest's tenure of office.
46. *CIL* X. 3698 = *ILS* 4175: M. Magrio Basso L. Ragonio | Quintiano cos. k. Iunis, Cumis in templo divi Vespa- | siani in ordine decurionum, | quem M. Mallonius Undanus | et Q. Claudius Acilianus praet. | coegerant, scribundo sorte | ducti adfuerunt Caelius Pan- | nychus, Curtius Votivos, Considi- | us Felicianus, referentibus pr. | de sacerdote faciendo Matris | deae (sic) Baianae in locum Restituti

| sacerdotis defuncti, placuit uni- | versis Licinium Secundum | sacerdotem fieri. | XV(viri) sac. fac. pr(aetoribus) et magistratibus Cuman(is) sal(utem): | Cum ex epistula vestra cognove- | rimus creasse vos sacerdotem | Matris deum Licinium Secundum | in locum Claudi Restituti defunc- | [t]i, secundum voluntatem vestram | permisimus ei occavo et | corona, dumtaxat intra | fines coloniae vestrae, uti. | Optamus vos bene valere. | Pontius Gavius Maximus | promagistro subscripsi XVI kal. | Septembres, M. Umbrio Primo | T. Fl. Coeliano cos.

47. *CIL* XIII. 1751 = *ILS* 4131: taurobolio Matris d. m. ID., | quod factum est ex imperio Matris | deum, | pro salute imperatoris Caes. T. Aeli | Hadriani Antonini Aug. Pii p. p. | liberorumque eius, | et status coloniae Lugudun., | L. Aemilius Carpus IIIIIvir Aug., item | dendrophorus | (caput bovis) | vires excepit et a Vaticano trans- | tulit, ara et bucranium | suo inpendio consacravit, | sacerdote | Q. Sammio Secundo ab XV viris | occabo et corona exornato, | cui sanctissimus ordo Lugudunens. | perpetuitatem sacerdoti decrevit, | App. Annio Atilio Bradua, T. Clod. Vibio | Varo cos. | L. d. d. d. (In dextro latere) Cuius mesonyctium | factum est V id. Dec.
48. Similar phrases, indicating that a taurobolium was performed on the order of Magna Mater, occur in other inscriptions: *CIL* II. 5521, IX. 3015, X. 1596, XII. 4321, 4323, 4325, and XIII. 11,352. These expressions could refer either to a religious experience on the part of the person who performed the taurobolium, or it could indicate that the taurobolium was prompted by a portent or by a dream experienced by the local priest or archigallus of the cult. The expression *ex vaticinatione archigalli* is discussed below. Book XI of Apuleius' *Metamorphosis* contains numerous instances in which Isis appears in dreams to issue commands.
49. Julian *Oratio* V.173D, 177A, Arnobius V.16, and Tertullian *De Ieiunio* 16.
50. For a brief but clear statement of Augustus' constitutional position see Jones 1951. For a detailed treatment of Augustus' assumption of tribunician power see Lacey 1996 110–116 and 154–168.
51. For detailed discussion of the emperor's so-called tribunician day see Hammond 1938, 1949, and 1959 72–76.
52. The three inscriptions containing calendrical dates that are not discussed here are: *CIL* XII. 11,567 (September 30, 245 A.D.), XIII. 1752 (June 16, 190 A.D.), and *AE* 1910 #217 and 1924 #26 (. . . III kal. . . . 245 A.D.).
53. Given the close verbal similarity between this document and two other texts (*CIL* XIII. 1752–1753 = *ILS* 4132–4133), there can be no doubt about this inscription being the record of a taurobolium of imperial type.
54. *CIL* XII. 1782 = *ILS* 4130: . . . | . . . domusq(ue) divi- | nae, Colon(iae) Copiae Claud(iae) Aug(ustae) Lug(dunensium) | taurobolium fecit Q. Aquius Antonia- | nus pontif. perpetuus | ex vaticinatione Pusoni Iuliani archi- | galli inchoatum XII kal. Mai., consum- |matum VIIII kal. Mai. L. Eggio Marullo | Cn. Papirio Aelio cos., praeeunte Aelio | C[astren]s[e] sacerdote, tibicine Albio | Verino.
55. Dio LXXIII.15–17, Herodian II.12, *Hist. Aug. Did. Iul.* 8–9, and *Severus* 5–6. For a modern acount of these events see Birley 1989 89ff.
56. Alternatively, it is possible that for the sake of brevity the phrase *pro salute*, followed by the lengthy official imperial titulature, was omitted from these texts, and that their imperial character would have been obvious to the ancient observer due to their physical relationship to *CIL* XIII. 511.
57. *CIL* XII. 4321 = *ILS* 4111: Matri | Deum | taurobolium indictum | iussu ipsius ex stipe conlata | celebrarunt publice Narbon(enses).
58. *CIL* XII. 4323 = *ILS* 4120: Imperio D(eum) M(atris) tauro- | polium provinciae | Narbonensis | factum per C. Batonium | Primum flaminem Aug(ustalem) | pro salute dominorum | Imp. L. Septimi Severi | Pii Pertinacis Aug. Ara- bici Adiabenici Parthi- | ci Maximi, et M. Aureli | Ant(onini) Aug. . . .

59. *CIL* VIII. 8203 = 19,981 (= *ILS* 4136), XII. 1782 (= *ILS* 4130), and XIII. 1752 (= *ILS* 4132).
60. *Fragmenta Vaticana* s.148 = S. Riccobono, *FIRA* II. p.496: "Item is, qui in Portu pro salute imperatoris sacrum facit ex vaticinatione archigalli, a tutelis excusatur."
61. Cicero was born on January 3 of 106 B.C. (Cic. *ad Att.* XIII.42.2, and Plut. *Cic.* 2.1). Plutarch states that Pompey, who was also born in 106 B.C., died on the day after his birthday (*Pompey* 79.4). The *Feriale Duranum* shows that Julius Caesar's birthday was still being commemorated on July 12 by the Roman army during the reign of Severus Alexander. For the custom of celebrating birthdays from the Augustan to the Flavian period see Horace *Odes* IV.11, Tibullus II.14 (IV.8), Ovid *Tristia* V.5.1–12, Pliny *Ep.* III.7.3, Martial X.24, 29, and 87.
62. Caligula was born on Aug. 31 (Suet. *Cal.* 8.1), and Nero was born on December 15 (Suet. *Nero* 6.1). For the birth dates of L. Verus and Commodus and their identification with these two earlier emperors see *Hist. Aug. L. Veri* 1.8, 4.6, 10.8 and *Commodi* 1.1, 10.2.
63. See, for example, *ILS* 7196, 8370, 8374, 8376, and 8379.
64. *AE* 1956 #255: M(atri) D(eum) s(acrum) | duo Irinaei, pater et | fil(ius), criobolati | natali suo, sacer(dotibus) | Lucio Antist(io) Avito | G. Antisti(o) Felicis- | simo.

    The words *natali suo* need not mean that the father and son shared the same birthday. The words should more properly be taken in a distributive sense. The two priests mentioned here could have supervised both ceremonies within the same year. Alternatively, the two ceremonies might not have occurred in the same year, and one priest supervised the criobolium of the father, and the other did so for the son.
65. *CIL* II. 5260 = *ILS* 4156: M(atri) D(eum) s(acrum) | Val(eria) Avita | aram tauroboli | sui natalici red- | diti d(edit) d(edicavit), sacerdo- | te Doccyrico Vale- | riano, arc<h>igallo | Publicio Mystico.

    The precise meaning of *redditi* is not absolutely clear, but it could mean 'renewed' in the sense that Valeria Avita's observance of the taurobolium on her birthday also made the same day the date of her mystic rebirth. This text has been dated to the latter portion of the second century A.D. on the basis of its letter forms.
66. *CIL* XIII. 573 = *ILS* 4157: natalici virib(us) | Valer(ia) Iullina | et Iul(ia) Sancta.
67. For the ancient astrological signs of the twelve months see the Menologium Rusticum Colotianum = *ILS* 8745. This inscribed marble altar associates Aries with April and Taurus with May. Ovid dates the rising of Aries to March 23, the rising of Taurus to April 20, and the rising of Gemini to May 20 (*Fasti* III.851–870, IV.713–720, and V.693–696).
68. For a striking illustration of how astrological beliefs influenced the Emperor Augustus see Pliny *NH* II.93–94 with Ramsey and Licht 1997 147–159.
69. Livy I.45 and Dion. Hal. IV.25–26. For the temple's *dies natalis* see the *Fasti Antiates Maiores* = *ILLRP* 9.
70. Line 62 of the anonymous *Carmen Contra Paganos* ridicules the taurobolium by suggesting that it was believed to confer twenty years of life upon the celebrant: "vivere num speras viginti mundus in anos?" The testimony of this blatantly anti-pagan work cannot be accepted at face value. In fact, it is better to regard this statement as a deliberate distortion designed to deny categorically the pagan claim that the taurobolium possessed the same mystical efficacy as Christian baptism. For this comparison and contrast see Firmicus Maternus *De Errore Profanarum Religionum* 27.8.

71. The *Fasti Ostienses* (Degrassi 1963 203 and 233) record the celebration of Hadrian's *ludi votivi decennales* on October 20–30 of 127. The adjective 'votivi' suggests that the games were vowed at the time of Hadrian's accession. The acts of the Arval Brethren record decennial vows at the accessions of Pertinax in 193, for Elagabalus in 218, and for Gordian III in 239.Hadrian's decennalia is also recorded in a papyrus from Egypt. See S. Eitrem and Leiv Amundsen, *Papyri Osloenses*, Oslo 1936, vol. III. p. 45 no. 77 col. ii. 1115–1116.
72. In fact, *CIL* XIII. 520 (= *ILS* 4125) is a fragmentary inscription from a taurobolic altar recording the performance of the ritual *pro salute et incolumitate* for an emperor or imperial personages whose identity or identities are unknown.
73. For these events and their dates see *Hist. Aug. Marci* 16.2, 17.3 and *Commodi* 2.2–4, 12.4. For modern treatment of these events see Birley 1987 184–189 and 195–197.

NOTES TO CHAPTER 6

1. For other modern treatments of this complex subject see Usener 1905, 1911, Cumont 1911, Duchesne 1919, Norden 1924, Botte 1932, Noiville 1936, Lake 1910, 1912, 1937, Strittmatter 1942, McArthur 1953 31–76, Bainton 1962, Culson 1962 18–36, and Kraabel 1982.
2. The author follows the generally accepted scholarly view of the synoptic Gospels, established by German scholars of the nineteenth century, that *Mark* was the first of the Gospels to be written, and that *Matthew* and *Luke* independently reworked *Mark* and supplemented its content with other material.
3. For an exhaustive analysis of these two nativity narratives see Brown 1993.
4. See, *inter alia*, Ramsay 1917, Taylor 1933, Braunert 1957, Sherwin-White 1963 162–171, Finegan 1964 234–238, and Brown 1993 547–556 and 666–668.
5. An inscription of unknown provenience (*ILS* 2683) records the career of a certain Q. Aemilius Secundus. Besides listing the offices that he held in his community following his military service, it records his service as prefect of a cohort under Quirinius in Syria, and it further indicates that he was given the task of carrying out a census of Apamea, resulting in the enumeration of 117,000 citizens. Although the inscription provides no date for the latter event, it does offer independent confirmation of Quirinius having supervised a census of Syria.
6. *ILS* 918: ". . . . . r]egem, qua redacta in pot[estatem Imp. Caesaris] | Augusti populique Romani [s]enatus [dis immortalibus] | supplicationes binas ob res prosp[ere gestas, et] | ipsi ornamenta triumph[alia decrevit;] | pro consul. Asiam provinciam op[tinuit; legatus pr. pr.] | Divi Augusti iterum Syriam et Ph[oenicen potinuit."
7. See Taylor 1933 126, Broughton 1933, Syme 1934 131ff, and Levick 1967 212.
8. See Groag "P. Sulpicius (90) Quirinius" in *RE* IV.1 (1931) 822–843, Taylor 1936, and Syme 1972.
9. See *Mark* 2.18ff, *Matthew* 9.14ff, 11.2ff, and *Luke* 7.11ff. Modern scholars generally regard the story of John's baptism of Jesus as a Christian tale designed to deny the Baptist's independence from Jesus. See *Mark* 1.1–11, *Matthew* 3, *Luke* 3.15–22, *John* 1.6–36, and 3.25ff.
10. The *Fasti Amiternini* and the *Fasti Antiates Ministrorum Domus Augustae* (Degrassi 1963 193 and 209)record this day as the one on which the senate decreed Augustus' deification. Tacitus (*Ann.* I.10–1) indicates that it was during the same session of the senate that Tiberius was officially voted his powers to succeed Augustus.

11. The author reckons Augustus' accession to power from August 19 43 B.C. when he first entered upon the consulship.
12. This festival is recorded in an inscription dating to 239 B.C. . See *OGIS* 56 l.64. For further discussion of the festival as the model for Christian Epiphany see Norden 1924 14–50.
13. Pliny (XVIII.256), Columella (*De Re Rust.* XI.2.49), and the Calendar of Furius Philocalus of 354 A.D. (Degrassi 1963 249) all agree in assigning the summer solstice to June 24, whereas Ovid (*Fasti* VI.785–790) and the *Fasti Venusini* (Degrassi 1963 59) place it two days later on June 26.
14. For detailed treatment of the ancient evidence see Humphrey 1986 91–95.
15. Before leaving this topic,it should be pointed out that an alternative explanation can be given for August 9 being sacred to Sol Indiges. As mentioned at the beginning of this section, Varro (*De Re Rust.* I.28) placed the turning points of the solar cycle not at the beginnings of the four seasons, but at their midpoints. Thus, his dating for the beginnings of the four seasons are as follows: February 7 for spring, May 9 for summer, August 11 for autumn, and November 10 for winter. It is therefore possible that Roman priests chose August 9 to be the day of Sol Indiges, because it was regarded as signaling the end of summer and the beginning of autumn.
16. For more detailed treatments of Egyptian obelisks in general and of the ones in Rome discussed in the following pages see D'Onofrio 1965, Iverson 1968, and Richardson 1992 272–276 with Pliny XXXVI.64–74 and Amm. Marc. XVII.4.
17. For Caesar's borrowing the calendar from Egypt see Pliny XVIII.211, Appian *Bell. Civ.* II.154, Dio XLIII.26, and Macrob. *Sat.* I.14.3.
18. *ILS* 91: «Imp. Caesar divi f. | Augustus | pontifex maximus, | imp. XII, cos. XI, trib. pot. XIV, | Aegupto in potestatem | populi Romani redacta | Soli donum dedit.»
19. Pliny XXXVI.72: "Ei, qui est in Campo, divus Augustus addidit mirabilem usum ad deprendendas solis umbras dierumque ac noctium ita magnitudines, strato lapide ad longitudinem obelisci, cui par fieret umbra brumae confectae die sexta hora paulatimque per regulas, quae sunt ex aere inclusae, singulis diebus decresceret ac rursus augesceret."
20. *OGIS* 458. Copies of this inscribed decree have been found at several cities of the Roman province.
21. See Tertullian *Apologia* 6.8, *Ad Nationes* I.10.17–18, Dio XL.47.3–4, XLII.26.2, and Val. Max. I.3.4.
22. Tac. *Ann.* II.85, Suet. *Tib.* 36, and especially Josephus *Ant. Iud.* XVIII.3.4.
23. For an excellent treatment of the ancient evidence on this matter see Colson 1926.
24. Ovid *Ars Amatoria* I.75–76, 416, *Remedia Amoris* 219–220, and Horace Sat. I.9.67. Cf. Suet. *Tib.* 32, Seneca *Ep.* 95, Frontinus *Strat.* II.1.17, and Juvenal XIV.96.
25. "Mercurii" is no longer visible on the line that obviously bore the word, and what remains visible of "Veneris" are all the letters except for faint traces of the two e's.
26. The modern literature on the different components of the religious culture of the Roman Empire is vast and often highly specialized. In recent years volumes of *Aufstieg und Niedergang der römischen Welt* II. contain numerous articles that survey modern scholarship and provide extensive bibliography. For general surveys of the religions in Imperial times see Cumont 1929, Dodds 1965, Ferguson 1982, Martin 1987, Turcan 1996, and Beard-North-Price 1998.
27. For this phenomenon see Fauth 1995.

28. Hijmans (1996 115ff) has set forth the most systematic and powerful critique of the earlier modern consensus on this matter. In doing so he has drawn upon and further developed the work of Seyrig (1971) and Drijvers (1976 20). Dirven (1999 157–189) has accepted this new approach and has applied it to her own study of religious assimilation in reference to Palmyrenes who resided in Dura-Europos. Houston (1990) has offered a new interpretation of an altar from Rome, dating to the second century of our era, which had long been viewed as solid proof for the Syrian god Malakbel having played an important role in the formation of Sol Invictus.
29. See *LIMC* IV. 1 592–596 with 2 366–385.
30. For the Republican coinage see Crawford 1974 150 #39 (217–215 B.C.), 280 #250 (132 B.C.), 314 #303 (109–108 B.C.), 318 #309 (118–107 B.C.), 318–319 #310 (118–107 B.C.), 457–459 #437 (51 B.C.), 473–475 #463 (46 B.C.), 483–485 #474 45 B.C.), 502–511 #494 (42 B.C.), 512 #496 (42 B.C.), and 533 #534 (38 B.C.). For Vespasian see Mattingly-Sydenham 1926 18 #28. For Trajan see Mattingly-Sydenham 1926 267–268 #326–330 and #341–342, and 307 #785. For Hadrian see Mattingly-Sydenham 1926 340–341 #16 and #20, 345 #43, 357 #145, 360 #167–168, and 426 #661. For Commodus see Mattingly-Sydenham 1930 379 #119. For additional Imperial coinage see Hijmans 1996 135–138. For Sol Invictus on the coins of Constantine see Bruun 1958.
31. For a full collection of the epigraphic data and archaeological artifacts see Vermaseren 1956 and 1960. For a survey of scholarship with an extensive bibliography see Beck 1984. Hinnells 1975, Duchnes-Guillemin 1978, Bianchi 1979, and Hinnells 1994 are collections of papers by various scholars treating numerous aspects of Mithraism at international conferences. Ulansey 1989, Gordon 1996, Turcan 2000, Clauss 2000, and Beck 2006 are attempts to reconstruct Mithraism. Beck 1988 focuses upon Mithraism's use of astrological lore.
32. *ILS* 4190: "Invicto Mithrae | Apronianus arkar. rei P(ublicae) d(2onum) d(edit) | dedicatum VII K. Iul. | Maximo et Orfito cos. | per C. Arennium Rea- | tinum patrem."
33. For a full and detailed treatment of Sol Invictus in reference to the reigns of Elagabalus and Aurelian see the studies of Halsberghe 1972, 1984, and Turcan 1985. For Emesa in the larger context of the Roman Near East see Millar 1993 300–310.
34. For the dynasty of Emesa see Sullivan 1977, and for Julia Domna see Levick 2007.
35. Our principal ancient sources for the reign of Elagabalus are Dio LXXX.1–21, Book V of Herodian, and his biography in the Historia Augusta, whose first 18 chapters seems fairly reliable, after which the account becomes fictional. Halsberghe in his first chapter has conveniently collected and set forth the literary texts relevant to Sol Invictus from Varro to the end of antiquity, which, of course, includes the reigns of Elagabalus and Aurelian. For modern studies of the religious policy of Elagabalus see Pietrzykowski 1978 and Frey 1989.
36. This conical black stone, described by Herodian (V.3.5), is shown on coins of Elagabalus. See Mattingly-Sydenham 1938 41 #176.
37. For these structures see Richardson 1992 7, 82, and 261–262.
38. Herodian's wording, *akmazontos therous*, is unfortunately imprecise, but it could indicate that this ceremony occurred at the very time of the summer solstice.
39. For the history of this obelisk see D'Onofrio 1965 295–297 and Iversen 1968 161–173.
40. For a study of these years through the prism of the Thirteenth Sibylline Oracle see Potter 1990.

41. The surviving ancient literary sources for the history of this period are scanty, sketchy, and often unreliable and have to be supplemented with the evidence from coins and inscriptions, which have their own problems of interpretation, thus making the reconstruction of the events of these years very challenging for modern scholars. The best recent detailed treatment is to be found in the chapters of *The Cambridge Ancient History, Volume XII, The Crisis of the Empire, A.D. 193–337*, edited by A. K. Bowman, P. Garnsey, and A. Cameron, Cambridge 2005.
42. For detailed treatment of Decius' religious policy see Rives 1999.
43. For detailed treatments of this much discussed topic, the persecution of Christians in Roman Imperial times see, for example, Hardy 1910 29–161, Frend 1965, Keresztes 1979, Lane-Fox 1986 419–492, Sordi 1986, De Ste. Croix 2006 35–78, 105–152, and J. Streeter's historiographical essay in De Ste. Croix 2006 3–34.
44. For the surviving texts of these certificates see Knipfing 1923.
45. For surveys of Palmyra's history as a Roman Imperial trading center and the conflict between Zenobia and Aurelian see Michalowski 1960, Stoneman 1992, Millar 1993 159–173, and 319–36. For its religion and gods see Drijvers 1976, Teixidor 1979, and Gawlikowski 1990. For additional bibliography see Drijvers 1977 837–863. Although Dirven 1999 is primarily concerned with the Palmyrenes resident in Dura-Europos, the study includes much valuable information about Palmyra itself. See, for example, the brief but excellent survey on pp.17–29.
46. For full, detailed treatments of Aurelian's reign see Groag in *RE* V.1 1903 1347–1419 s.v. "Domitius (36) Aurelianus," Homo 1904, Christol 1994, and Watson 1999. Despite its length, the biography of Aurelian in the *Historia Augusta* is padded with much fictitious material, on which see Fisher 1929. Much briefer, more sober, and generally more reliable is Zosimus' account of Aurelian's reign in I.47–62. Sotgiu 1975 surveys modern scholarship with an extensive bibliography.
47. Aur. Victor *De Caes.* 35.7, Eutrop. IX.15.1, *Hist. Aug. Aurelian* 10.2, 28.5, 35.3, 39.6, *Tacitus* 9.2, and Zosimus I.61.2.
48. Macrobius (*Sat.* III.9.7–8) records the religious formula used in 146 B.C. in performing the *evocatio* of Juno from Carthage.
49. For Aurelian's coins see Mattingly-Sydenham 1927 248–312, and for the interpretation followed here see Watson 1999 183–191.
50. For detailed treatment of the period encompassed by the reigns of Diocletian and Constantine covered here see Barnes 1982 (a thematic and institutional study), Odahl 2003 (a historical narrative), and the relevant Chapters in *The Cambridge Ancient History, Volume XII, The Crisis of the Empire, A.D. 193–337*, 2005. For the religious climate of the period see Liebeschuetz 1979 223–252.
51. There was even a third and fourth usurpation, both by Maximian who had gone into retirement along with Diocletian, but then emerged twice again in an attempt to recover his status as emperor.
52. Liebeschuetz (1979 279) regards as historical this detail in Eusebius' account, but Hollerich (1989) has argued convincingly that it was the Christian bishop's own invention and was patterned after *Exodus* 3.6–15, where Moses asks the god of his father to identify himself. This detail was part of Eusebius' overall scheme to portray Constantine as the Christian equivalent of the Jewish Moses.
53. Lact. *De Mort. Pers.* 44.5–6: «Commonitus est in quiete Constantinus, ut caeleste signum Dei notaret in scutis atque proelium committeret. Fecit, ut iussus est, et transversa X littera, summa capite circumflexo, Christum in scutis notat. 6 Quo signo armatus, exercitus capit ferrum.»

54. For this historical problem see Eadie 1977, which conveniently surveys modern scholarly interpretations from Edward Gibbon to the 1960's. For more recent scholarship on Constantine and his Christianity see Dimaio-Zeuge-Potov 1988 and Cameron-Hall 1999.
55. E.g., Obsequens 14 and 20 for the years 163 and 147 B.C. Dimaio, Zeuge, and Potov (1988 341ff) have offered a different astronomical explanation for Constantine's vision, which combines elements of Lactantius' and Eusebius' versions. They contend that on the night before the battle Constantine and perhaps some of his soldiers beheld in the night sky an unusual conjunction of the planets Venus, Mars, Jupiter, and Saturn, which together with stars from the constellation Capricorn actually would have formed a Chi-Rho monogram.
56. Given Constantine's pivotal role in the Christianizing of the Roman Empire, the crucial passage of this speech (22) has attracted much modern scholarly attention, because it offers an unusual insight into Constantine's pre-Christian religiosity. For a discussion of the speech with citation of other modern studies see Rodgers 1980.
57. Pan. Lat. VI.21.3–7: "Postridie enim quam accepto illo nuntio geminatum itineris laborem susceperas, omnes fluctus resedisse, omnem quam reliqueras tranquillitatem redisse didicisti, ipsa hoc sic ordinante Fortuna ut te ibi rerum tuarum felicitas admoneret dis immortalibus ferre quae voveras, ubi deflexisses ad templum toto orbe pulcherrimum, immo ad praesentem, ut vidisti, deum. 4 Vidisti enim, credo, Constantine, Apollinem tuum comitante Victoria coronas tibi laureas offerentem, quae tricenum singulae ferunt omen annorum. Hic est enim humanarum numerus aetatum quae tibi utique debentur ultra Pyliam senectutem. 5 Et—immo quid dico "credo?"—vidisti teque in illius specie recognovisti, cui totius mundi regna deberi vatum carmina divina cecinerunt. 6 Quod ego nunc demum arbitror contigisse, cum tu sis, ut ille, iuvenis et laetus et salutifer et pulcherrimus imperator. 7 Merito igitur augustissima illa delubra tantis donariis honestasti.»
58. For similar arguments in favor of Constantine's worship of Sol Invictus see Liebeschuetz 1979 241 and 279–284.
59. For a list of the numerous legends and types for the emperors other than Constantine during the years 294–313 see Southerland 1967 704 and 714. For a similar list for the reigns of Constantine and Licinius see Bruun 1966 751–753.
60. *ILS* 659 = Vermaseren 1960 1698: "D(eo) S(oli) I(nvicto) M(ithrae) | fautori imperii sui | Iovii et Herculii | religiosissimi | Augusti et Caesares | sacrarium | restituerunt."
61. In 311 a Mithraeum in Noricum, which had collapsed and had been deserted for fifty years, was repaired (Vermaseren 1960 1431), and at about the same time in the same region a new Mithraeum was built (Vermaseren 1960 1414). At Poetovio in Upper Pannonia an equestrian commander restored a Mithraeum (Vermaseren 1960 1614), and in Lower Moesia another equestrian commander did likewise (Vermaseren 1960 2280). In 325 in Upper Germany a man belonging to the Mithraic grade of *Corax* built a new Mithraeum upon ground that formed part of his gift to the god, whose shrine was dedicated in honor of the imperial house (Vermaseren 1960 1313–1322).
62. Tertullian (*Adversus Iudaeos* 8), writing about a century before Lactantius, assigned Jesus' death to March 25, and he is the earliest Christian writer known to have done so. Like Lactantius, Tertullian was virulently antipagan and would have scoffed at the idea of the date of Jesus' death as resulting from his life's association with the solar cycle; and since, like Lactantius, Tertullian records this date in a matter-of-fact fashion as if it were common

knowledge, the tradition of this dating is likely to have preceded Tertullian by some time.

63. *Codex Iust.* III.12.2: «Omnes iudices urbanaeque plebes et artium officia cunctarum venerabili die Solis quiescant. Ruri tamen positi agrorum culturae libere licenterque inserviant, quoniam frequenter evenit, ut non alio aptius die frumenta sulcis aut vineae scrobibus commendentur, ne occasione momenti pereat commoditas caelesti provisione concessa.»
64. Anna Comnena *Alexiad* XII.4.5. For more detailed treatment of this statue see Preger 1901, Dimaio-Zeuge-Potov 1988 355–357, and Bassett 2004 201–204.
65. *ILS* 736: «Victor ovans urbique favens sublime tropaeum | principis et munus condit decoratque triumphis.»
66. For coercion as a factor in conversion see MacMullen 1984 86–101 and De Ste. Croix 2006 201–229. For other modern treatments of the conflict between Christianity and paganism during the fourth century see Huttman 1914, Piganiol 1947, King 1960 71–93, Momigliano 1963, Downey 1963 143–199, Wytzes 1977, Geffcken 1978 115ff, Liebeschuetz 1979 291ff, Croke-Harries 1982, Lane Fox 1986 663–680, Chuvin 1990 1–72, Kirsch 2004 119–284, Hahn 2004, and Salzman in Rüpke 2007 109–125.
67. Euseb. *HE* X.7 and *Cod. Theod.* XVI.2.1.
68. Thanks to the survival of the historical account of Ammianus Marcellinus, Julian's own writings, and the speeches of the contemporary sophist Libanius, Julian's reign is well documented and has received considerable modern scholarly attention: see the biographies of Browning 1976, Bowersock 1978, and Murdoch 2003.
69. For recent modern studies of so-called pagan monotheism in the philosophy and religion of the Later Roman Empire see Athanassiadi-Frede 1999.
70. For the Christian destruction of the famous temple of Serapis in Alexandria in 391 see the *Ecclesiastical Histories* of Socrates (V.15–6) and Sozomen (VII.15+20).
71. It is interesting to note that some sixty years later Leo, the bishop of Rome (Pope Leo the Great 440–481), revived *pontifex maximus* and adopted it as the formal Latin title for the pope, which it remains to this day.
72. I follow here the modern traditional historical interpretation of these events as being in part a conflict between paganism and Christianity, as portrayed by contemporary or near contemporary Christian historians. See Rufinus II.31–3, Socrates V.25, and Sozomen VII.22+24. . This modern orthodoxy, as best set forth by Bloch (1945), has, however, in recent decades been challenged by a revisionist interpretation, according to which paganism went out with an apathetic whimper rather than a combative bang. For this revisionist view see Cameron 1966, O'Donnell 1977, 1978, 1979, Cameron 1982, Salzman 1990 193–246, and especially Cameron 2011. For discussion of these two competing interpretations with a vindication of the traditional view see Hedrick 2000 39–68.
73. For the phenomenon of so-called pagan survivals see Green 1931, Hyde 1963, Laing 1963, Frantz 1965, Saradi-Mendelovici 1970, Nassivera 1976, Hanson 1978, Chuvin 1990 73–150, MacMullen 1997, and Hahn-Emmel-Gotter 2008.
74. The author disagrees with Duchesne (1919 258 and 290), who dated the calendar and its attendant material to the year 336. On this matter see Salzman 1990 279–282.
75. Our knowledge of the chronology of December 25th's adoption as Jesus' birthday throughout the Roman Empire derives from brief allusions to the

celebration of Christmas scattered throughout patristic literature. For the ancient evidence and modern scholarly inferences therefrom see Usener 1911 221–259, 329–347, and Strittmatter 1942 600–611.

76. For a detailed treatment of this calendar see Dulabahn 1986.
77. *Patrologia Latina* LIV. 218–219 = *Sermo* XXVII, *in Nativitate Domini* VII, 4, as translated by Strittmatter (1942 614).
78. For detailed treatment of these discoveries see Toynbee and Ward-Perkins 1956.

Bibliography

Alföldi, A. 1948: *The Conversion of Constantine and Pagan Rome*, tr. H. B. Mattingly, Oxford.

———. 1952: *A Conflict of Ideas in the Late Roman Empire: The Clash between the Senate and Valentinian I*, tr. H. B. Mattingly, Oxford.

———. 1960: "Diana Nemorensis," *AJA* 64 137–144.

———. 1965: *Early Rome and the Latins*, Ann Arbor MI.

———. 1977: "From the *Aion Plutonios* of the Ptolemies to the *Saeculum Frugiferum* of the Roman Emperors (Redeunt Saturnia Regna VI)," *Greece and the Eastern Mediterranean in Ancient History and Prehistory: Studies Presented to Fritz Schachermeyr on the Occasion of his Eightieth Birthday*, ed. K. H. Kinzl, Berlin-New York 1–30.

Allély, A. 2003: "Les infants malformés et Considérés comme 'Prodigia' à Rome et en Italie sous la République," *REA* 105 127–156.

Allen, B. M. 1947/8: "The Early Roman Calendar," *CJ* 43 163–168.

Altheim, F. 1938: *A History of Roman Religion*, tr. H. B. Mattingly, London.

Ampolo, C. 1970: "L'Artemide di Marsiglia e la Diana dell'Aventino," *PdP* 25 200–210.

Ando, C. (editor). 2003: *Roman Religion*, Edinburgh.

Athanassiadi, P., and M. Frede (editors). 1999: *Pagan Monotheism in Late Antiquity*, Oxford.

Bagnani, G. 1919: "The Subterranean Basilica at Porta Maggiore," *JRS* 9 78–85.

Bainton, R. H. 1962: "The Origins of Epiphany," *Collected Papers in Church History*, Boston, I. 22–38.

Balsdon, J. P. V. D. 1966: "The *Salii* and Campaigning in March and October," *CR* N.S. 16 146–147.

Barnes T. D. 1981: *Constantine and Eusebius*, Oxford.

———. 1982: *The New Empire of Diocletian and Constantine*, Cambridge MA.

———. 1985: "The Conversion of Constantine," *Echos du Monde Classique* N.S. 4 371–391.

———. 1989: "Christians and Pagans in the Reign of Constantius," *L'Église et L'Empire au IVe Siècle, Entretiens sur L'Antiquité Classique, Tome XXXIV*, Vandoeuvres-Genève, 301–337.

Barton, T. 1995: "Augustus and Capricorn: Astrological Polyvalency and Imperial Rhetoric," *JRS* 85 33–51.

Bassett, S. 2004: *The Urban Image of Late Antique Constantinople*, Cambridge.

Bayet, J. 1926: *Les Origines de l'Hercule Romain*, Paris.

———. 1969: *Histoire politique et psychologique de la Religion Romaine*, Paris.

Beard, M. 1980: "The Sexual Status of Vestal Virgins," *JRS* 70 12–27.

———. 1985: "Writing and Ritual: A Study in Diversity and Expansion in the Arval *Acta*," *PBSR* 53 114–162.

———, J. A. North, and S. Price. 1998: *Religions of Rome, Volume I, A History*, Cambridge.
Beck, R. 1984: "Mithraism since Franz Cumont," *ANRW* II. 17.4 2002–2114.
———. 1988: *Planetary Gods and Planetary Orders in the Mysteries of Mithras*, Leiden.
———. 1998: "The Mysteries of Mithras: A New Account of their Genesis," *JRS* 88 115–128.
———. 2006: *The Religion of the Mithras Cult in the Roman Empire: Mysteries of the Unconquered Sun*, Oxford.
Bernstein, F. 1998: *Ludi Publici: Untersuchungen zur Entstehung und Entwicklung der Öffentlichen Siele im republikanischen Rom*, Stuttgart.
Bianchi, E. 2010: *Il Rex Sacrorum a Roma e nell'Italia Antica*, Milan.
Bianchi, U. 1979: *Mysteria Mithrae: Atti del Seminario internazionale sulla Specificità storico-religiosa di Misteri di Mithra, Con Particolare Riferimento alle Fonti documentari di Roma e Ostia, Roma e Ostia, 28–31 Mar.*, Leiden.
Bickerman, E. J. 1980: *Chronology of the Ancient World*, Ithaca NY.
Birley, A. R. 1987: *Marcus Aurelius: A Biography*, New Haven CT.
———. 1989: *Septimius Severus: The African Emperor*, New Haven CT.
Bispham, E., and C. J. Smith (editors). 2000: *Religion in Archaic and Republican Rome and Italy: Evidence and Experience*, Edinburgh.
Bloch, H. 1945: "A New Document of the Last Pagan Revival in the West, 393–394 A.D.," *HTR* 38 199–244.
Bloch, R. 1960: *The Origins of Rome*, London.
Boardman, J. 1980: *The Greeks Overseas*, London.
Bömer, F. 1957–8: *P. Ovidius Naso: Die Fasten*, Two Volumes, Heidelberg.
———. 1964: "Claudia Quinta," *MDAI* 71 130–151.
Botte, B. 1932: *Les Origines de la Noel et de l'Epiphanie*, Berlin-Leipzig.
Bowden, H. 2010: *Mystery Cults of the Ancient World*, Princeton NJ.
Bowersock, G. W. 1978: *Julian the Apostate*, London.
Bowman, A. K., and J. D. Thomas. 1994: *The Vindolanda Writing-Tablets (Tabulae Vindolandenses II)*, London.
Boyce, A. A. 1937: "The Expiatory Rites of 207 B.C.," *TAPA* 68 157–171.
Boyce, G. K. 1937: *Corpus of the Lararia of Pompeii*, Rome.
Brauer, G. C. 1975: *The Age of the Soldier Emperors: Imperial Rome, A.D. 244–284*, Park Ridge NJ.
Braunert, H. 1957: "Der römische Provinzialzensus und der Schätzungsbericht des Lukas-Evangeliums," *Historia* 6 192–214.
Bremmer, J. N. 1982: "The *Suodales* of Poplios Valesios," *ZPE* 47 133–147.
———. 1983: "Scapegoat Rituals in Ancient Greece," *HSCP* 87 299–320.
———. 1993: "Three Roman Aetiological Myths," *Mythos in mythenloser Gesellschaft: Das Paradigma Roms*, ed. F. Graf, Colloquium Rauricum 3, Stuttgart and Leipzig, 158–174.
Brind'Amour, P. 1978: "L'Origin des Jeux Séculaires," *ANRW* II. 16.2 1334–1417.
Brown, P. 1996: *The Rise of Western Christendom: Triumph and Diversity A.D. 200–1000*, Oxford.
Brown, R. E. 1993: *The Birth of the Messiah: A Commentary on the Infancy Narratives in Matthew and Luke*, Second Edition, Garden City NY.
Browning, R. 1976: *The Emperor Julian*, Berkeley CA.
Bruun, P. 1958: "The Disappearance of Sol from the Coins of Constantine," *Arctos* N.S. 2 15–37.
———. 1962: "The Christian Signs on the Coins of Constantine," *Arctos* N.S. 3 5–35.
———. 1966: *The Roman Imperial Coinage, Volume VII, Constantine to Licinius, A.D. 306–337*, London.

Buchner, E. 1982: *Die Sonnenuhr des Augustus*, Mainz.
Burkert, W. 1987: *Ancient Mystery Cults*, Cambridge MA.
Cameron, Alan. 1966: "The Date and Identity of Macrobius," *JRS* 56 25–38.
———. 1968: "Gratian's Repudiation of the Pontifical Robe," *JRS* 58 96–102.
———. 1977: "Paganism and Literature in Late Fourth Century Rome," *Christianisme et Formes Littéraires de L'Antiquité tardive en Occident, Entretiens sur L'Antiquité Classique, Tome XXIII*, Vandoeuvres-Genève, 1–30.
———. 2011: *The Last Pagans of Rome*, Oxford.
Cameron, Averil, and S. G. Hall. 1999: *Eusebius, Life of Constantine: Introduction, Translation, and Commentary*, Oxford.
Carcopino, J. 1926: *La Basilique Pythagoricienne de la Porte Majeure*, Paris.
———. 1941: *Aspects mystiques de la Rome paienne*, Paris.
Chadwick, H. 2001: *The Church in Ancient Society from Galilee to Gregory the Great*, Oxford.
Christol, M. 1994: *L'Empereur Aurelian et son Temps*, Paris.
Chuvin, P. 1990: *A Chronicle of the Last Pagans*, tr. B. A. Archer, Cambridge MA.
Clauss, M. 1990: "Sol Invictus Mithras," *Athenaeum* 78 423–450.
———. 2000: *The Roman Cult of Mithras: The God and his Mysteries*, tr. Richard Gordon, New York.
Clerici, L. 1942: "Die *Argei*," *Hermes* 77 89–100.
Coarelli, F. 1993: "Note sui *Ludi Saeculares*," *Spectacles sportifs et scèniques dans le Monde étrusco-italique*, College de l'École française de Rome 172, 2211–2245.
Colin, J. 1954: "Les Consuls du César-Pharaon Caligula et l'Héritage de Germanicus," Latomus 13 394–416.
Colson, F. H. 1926: *The Week: An Essay on the Origin and Development of the Seven-Day Cycle*, Cambridge.
Corbishley, T., S.J. 1934: "A Note on the Date of the Syrian Governorship of M. Titius," *JRS* 24 43–49.
Cornell, T. J. 1991: "The Tyranny of the Evidence: A Discussion of the Possible Uses of Literacy in Etruria and Latium in the Archaic Age," *Literacy in the Roman World*, Journal of Roman Archaeology Suppl. 3, M. Beard et al, Ann Arbor 7–33.
Crawford, M. H. 1974: *Roman Republican Coinage*, Two Volumes, Cambridge.
Cristofani, M. 1972: "Sull'Origine e la Diffusione dell'Alfabeto Etrusco," *ANRW* I. 2 466–489.
Croke, B., and J. Harries (editors). 1982: *Religious Conflict in Fourth-Century Rome: A Documentary Study*, Sydney.
Culson, O. 1966: *The Early Church: Studies in Early Christian History and Theology*, Philadelphia PA.
Cumont, F. 1911: *La Natalis Invictik*, Paris.
———. 1929: *Les Religions orientales dans le Paganisme romain*, Paris.
Curran, J. R. 2000: *Pagan City and Christian Capital: Rome in the Fourth Century*, Oxford.
Degrassi, A. 1963: *Inscriptiones Italiae, XIII. 2, Fasti Anni Numani et Iuliani*, Rome.
De Ligt, L. 1989: *Fairs and Markets in the Roman Empire: Economic and Social Aspects of Periodic Trade in a Pre-Industrial Society*, Amsterdam.
De Ste. Croix, G. E. M. 2006: *Christian Persecution, Martyrdom, and Orthodoxy*, Oxford.
Dimaio, M.; J. Zeuge; and N. Potov. 1988: "*Ambiguitas Constantiniana*: The *Caeleste Signum Dei* of Constantine the Great," *Byzantion* 58 333–60.
Dirven, L. 1999: *The Palmyrenes of Dura-Europos: A Study of Religious Interaction in Roman Syria*, Leiden.
Dodds, E. R. 1965: *Pagan and Christian in an Age of Anxiety: Some Aspects of Religious Experience from Marcus Aurelius to Constantine*, Cambridge.

D'Onofrio, C. 1965: *Gli Obelischi di Roma*, Rome.
Downey, G. 1963: *Ancient Antioch*, Princeton NJ.
Drews, R. C. 1981: "The Coming of the City to Central Italy," *AJAH* 6 133–165.
Drijvers, H. J. W. 1976: *The Religion of Palmyra*, Leiden.
———. 1977: "Hatra, Palmyra, und Edessa: Die Städte der syrisch-mesopotamischen Wüste in der politischen, kulturgeschichtlicher und religionsgeschichtlicher Beleuchtung," *ANRW* II. 8 799–906.
Duchesne, L. 1919: *Christian Worship: Its Origin and Evolution, A Study of the Latin Liturgy up to the Time of Charlemagne*, Fifth Edition, tr. M. L. McClure, London.
Duchnes-Guillemin, J. (editor). 1978: *Études Mithriaques: Actes du II Congres international, Teheran*, Acta Iranica, Series 1, 4, Leiden.
Dulabahn, E. S. 1986: *Studies on the Laterculus of Polemius Silvius*, Unpublished Dissertation, Bryn Mawr College PA.
Dumézil, G. 1966: *Archaic Roman Religion*, tr. P. Krapp, Chicago IL.
Dunbabin, T. J. 1957: *The Greeks and Their Eastern Neighbors: Studies in the Relations Between Greece and the Countries of the Near East in the 8th and 7th Centuries B.C.*, London.
Du Quesnay, I. M. Le M. 1977: "Vergil's Fourth Eclogue," *Papers of the Liverpool Latin Seminar 1976*, ed. E. Cairns, Arca 2 Liverpool, 25–99.
Duthoy, R. 1969: *The Taurobolium: Its Evolution and Terminology*, Leiden.
Eadie, J. W. 1977: *The Conversion of Constantine*, Huntington NY.
Eckhardt, K. 1910: "Die armenischen Feldzüge des Lukullus: Pt. II Das Kriegsjahr 69," *Klio* 10 72–115.
Eckstein, A. M. 1982: "Human Sacrifice and Fear of Military Disaster in Republican Rome," *AJAH* 7 69–95.
Edelstein, E. J., and L. 1945: *Asclepius: Collection and Interpretation of the Testimonies*, Baltimore MA.
Fantham, E. 1998: *Ovid, Fasti, Book IV*, Cambridge.
Fauth, W. 1978: "Römische Religion im Spiegel der *Fasti* des Ovid," *ANRW* II. 16.1 104–186.
———. 1995: *Helios Megistos: Zur synkretistischen Theologie der Spätantike*, Leiden.
Feeney, D. C. 1998: *Literature and Religion at Rome: Culture, Contexts, and Beliefs*, Cambridge.
———. 2007: *Caesar's Calendar: Ancient Time and the Beginnings of History*, Berkeley CA.
Ferguson, J. 1982: *The Religions of the Roman Empire*, Second Edition, London.
Finegan, J. 1964: *Handbook of Biblical Chronology: Principles of Time Reckoning in the Ancient World and Problems of Chronology in the Bible*, Princeton NJ.
Fink, R. O.; Hoey, A. S.; and Snyder, W. F. 1940: "*Feriale Duranum*," *YCS* 7 1–222.
Fisher, W. H. 1929: "The Agustan *Vita Aureliani*," *JRS* 19 125–149.
Fishwick, D. 1966: "The Cannophori and the March Festival of Magna Mater," *TAPA* 97 193–202.
Forsythe, G. 1994: *The Historian L. Calpurnius Piso Frugi and the Roman Annalistic Tradition*, Lanham MD.
———. 2005: *A Critical History of Early Rome: From Prehistory to the First Punic War*, Berkeley CA.
———. 2007: "Claudius Quadrigarius and Livy's Second Pentad," *Companion to Greek and Roman Historiography*, ed. J. Marincola, Volume 2, 391–396.
Fowler, W. Warde. 1899: *The Roman Festivals of the Period of the Republic, An Introduction to the Study of the Religion of the Romans*, Oxford.
———. 1902A: "Doctor Wissowa on the *Argei*," *CR* 16 115–119.

———. 1902B: "The Number Twenty-Seven in Roman Ritual," *CR* 16 211–212.
———. 1911: *The Religious Experience of the Roman People*, Oxford.
———. 1912: "Mundus Patet 24th August, 5th October, 8th November," *JRS* 2 25–33.
Frantz, A. 1965: "From Paganism to Christianity in the Temples of Athens," *Dumbarton Oaks Papers* 19 187–205.
Frazer, Sir J. G. 1929: *Publii Ovidii Nasonis Fastorum Libri Sex, The Fasti of Ovid*, Five Volumes, edited with a translation and commentary, London.
Frend, W. H. C. 1965: *Martyrdom and Persecution in the Early Church: A Study of a Conflict from the Maccabees to Donatus*, Oxford.
Frey, M. 1989: *Untersuchungen zur Religion und zur Religionspolitik des Kaisers Elagabal*, Stuttgart.
Galinsky, K. 1996: *Augustan Culture: An Interpretive Introduction*, Princeton NJ.
Gargola, D. 1995: *Lands, Laws, Gods: Public Lands in Republican Rome*, Chapel Hill NC.
Gaster, T. H. 1961: *Thespis: Ritual, Myth, and Drama in the Ancient Near East*, New York.
Gawlikowski, M. 1990: "Les Dieux de Palmyre," *ANRW* II. 18.4 2605–2658.
Geffcken, J. 1978: *The Last Days of Greco-Roman Paganism*, tr. Sabine MacCormack, Amsterdam.
Giacobello, F. 2008: *Larari Pompeiani: Iconografia e Culto dei Lari in Ambito Domestico*, Milan.
Goodhue, N. 1975: *The Lucus Furrinae and the Syrian Sanctuary on the Janiculum*, Amsterdam.
Gordon, A. E. 1932: "On the Origin of Diana," *TAPA* 63 177–192.
———. 1934: *The Cults of Aricia*, Berkeley CA.
———. 1983: *Illustrated Introduction to Latin Epigraphy*, Berkeley CA.
Gordon, R. L. 1996: *Image and Value in the Graeco-Roman World: Studies in Mithraism and Religious Art*, Aldershop UK.
Gradel, I. 2002: *Emperor Worship and Roman Religion*, Oxford.
Graf, F. 1997: *Der Lauf des rollenden Jahres: Zeit und Kalender in Rom*, Stuttgart and Leipzig.
Graillot, H. 1912: *Le Culte de Cybèle, Mère des Dieux, à Rome et dans l'Empire romain*, Paris.
Green, S. J. 2004: *Ovid, Fasti I: A Commentary*, Leiden.
Green, W. M. 1931: "The Lupercalia in the Fifth Century," *CP* 26 60–69.
Grégoire, H. 1930: "La Conversion de Constantin," *Revue de l'Université de Bruxelles* 36 231–272.
Groth, A. 1929: "Der Argeerkultus," *Klio* 22 303–334.
Gruen, E. S. 1990: *Studies in Greek Culture and Roman Policy*, Leiden.
Hadzsits, G. D. 1930: "The Date of the Megalensian Games," *TAPA* 61 165–174.
Hahn, J. 2004: *Gewalt und religiöser Konflikt: Studien zu den Ausandersetzungen zwischen Christen, Heiden und Juden im Osten des römischen Reiches (von Konstantin bis Theodosius II)*, Klio Beiheft N.F. 8, Berlin.
Hahn, J.; S. Emmel; and U. Gotter (editors). 2008: *From Temple to Church: Destruction and Renewal of Local Cultic Topography in Late Antiquity*, Leiden.
Hajjar, Y. 1977: *La Triade de Héliopolis-Baalbek: Son Culte et sa Diffusion à Trovers les Textes Literaires et les Documents iconographiques et épigraphiques*, Two Volumes, Leiden.
Hallett, J. P. 1970: "Over Troubled Waters: The Meaning of the Title *Pontifex*," *TAPA* 101 219–228.
Halsberghe, G. H. 1972: *The Cult of Sol Invictus*, Leiden.
———. 1984: "Le Culte de *Deus Sol Invictus* à Rome au III-ème Siècleapès J. C.," *ANRW* II. 17.4 2181–2201.

Hammond, M. 1938: "The Tribunician Day During the Early Empire," *MAAR* 15 23–61.
———. 1949: "The Tribunician Day from Domitian through Antoninus," *MAAR* 19 37–76.
———. 1959: *The Antonine Monarchy*, Papers and Monographs of the American Academy in Rome vol. 19, Rome.
Hannah, R. 2005: *Greek and Roman Calendars: Constructions of Time in the Classical World*, London.
Hanson, R. P. C. 1978: "The Transformation of Pagan Temples into Churches in the Early Christian Centuries," *JSS* 23 257–267.
Hardy, E. G. 1910: *Studies in Roman History*, New York.
Harmon, D. P. 1978A: "The Public Festivals of Rome," *ANRW* II. 16.2 1440–1468.
———. 1978B: "The Family Festivals of Rome," *ANRW* II. 16.2 1592–1603.
Hatt, J. J. 1950: "La Vision de Constantin au Sanctuaire de Grand et l'Origine Celtique du *Labarum*," *Latomus* 9 427–436.
Healey, J. F. 1990: *The Early Alphabet*, vol. 9 in the series Reading the Past, Berkeley CA.
Hedrick, C. W., Jr. 2000: *History and Silence: Purge and Rehabilitation in Late Antiquity*, Austin TX.
Helgeland, J. 1978: "Roman Army Religion," *ANRW* II. 16.2 1470–1505.
Henzen, G. 1874: *Fratres Arvales: Acta Fratrum Arvalium Quae Supersunt*, Berlin.
Hepding, H. 1903: *Attis: Seine Mythen und sein Kult*, Giessen.
Herbert-Brown, G. 1994: *Ovid and the Fasti: An Historical Study*, Oxford.
———(editor). 2002: *Ovid's Fasti: Historical Readings at its Bimilennium*, Oxford.
Hijmans, S. E. 1996: "The Sun Which Did Not Rise in the East: The Cult of Sol Invictus in Light of Non-Literary Evidence," *Bulletin Antieke Beschaving* 71 115–150.
Hinnells, J. R. (editor). 1975: *Mithraic Studies: Proceedings of the First International Congress of Mithraic Studies*, Two Volumes, Manchester.
———(editor). 1994: *Studies in Mithraism: Papers Associated with the Mithraic Panel [at] the XVIth Congress of the IAHR, Rome 1990*, Rome.
Hirschfeld, O. 1881: "Das Neujahr des tribunicischen Kaiserjahres," *Wiener Studien* 3 97–108.
Hoenigswald, H. M. 1941: "On Etruscan and Latin Month-Names," *AJP* 62 199–206.
Holladay, A. J., and M. D. Goodman. 1986: "Religious Scruples in Ancient Warfare," *CQ* N.S. 36 151–171.
Holland, L. A. 1961: *Janus and the Bridge*, Papers and Monographs of the American Academy in Rome vol. 21, Rome.
Hollerich, M. 1989: "The Comparison of Moses and Constantine in Eusebius of Caesarea's *Life of Constantine*," *Studia Patristica* 19 80–95.
Homo, L. 1904: *Essai sur le Règne de l'Empereur Aurelian (270–275)*, Paris.
Hörig, m. 1983: "Iupiter Dolichenus," *ANRW* II. 17.4 2135–2179.
———, and E. Schwertheim. 1987: *Corpus Cultus Iovis Dolicheni*, Leiden.
Houston, G. W. 1990: "The Altar from Rome with Inscriptions to Sol and Malakbel," *Syria* 67 189–193.
Hughes, D. 1979: *The Star of Bethlehem: An Astronomer's Confirmation*, New York.
Humphrey, J. H. 1986: *Roman Circuses: Arenas for Chariot Racing*, Berkeley CA.
Huttman, M. A. 1914: *The Establishment of Christianity and the Proscription of Paganism*, New York.
Hyde, W. W. 1963: *Greek Religion and its Survivals*, New York.
Iversen, E. 1968: *Obelisks in Exile, Volume I, The Obelisks of Rome*, Copenhagen.
Jones, A. H. M. 1951: "The *Imperium* of Augustus," *JRS* 41 112–119.

———. 1962: *Constantine and the Conversion of Europe*, London.

Jordan, H. 1871: *Topographie der Stadt Rom im Alterthum*, I.2, Berlin.

Keaveney, A. 1992: *Lucullus, A Life*, London and New York.

Keresztes, P. 1979: "The Imperial Roman Government and the Christian Church I: From Nero to the Severi," *ANRW* II.23.1 247–315.

Kidger, M. R. 1999: *The Star of Bethlehem: An Astronomer's View*, Princeton NJ.

King, N. Q. 1960: *The Emperor Theodosius and the Establishment of Christianity*, Philadelphia PA.

Kirsch, J. 2004: *God Against the Gods: The History of the War Between Monotheism and Polytheism*, New York.

Knipfing, J. 1923: "The *Libelli* of the Decian Persecution," *HTR* 16 345–390.

König, A. and I. 1991: *Der römische Festkalender der Republik: Feste, Organisation, und Priesterschaften*, Stuttgart.

Köves, Th. 1963: "Zum Empfang der Magna Mater im Rom," *Historia* 12 321–347.

Kraabel, A. T. 1982: "The Roots of Christmas," *Dialog* 21 274–280.

Lacey, W. K. 1996: *Augustus and the Principate: The Evolution of the System*, Leeds.

Laing, G. J. 1963: *Survivals of Roman Religion*, New York.

Lake, K. 1910: "Christmas," *Hastings Encyclopaedia of Religion and Ethics*, III. 601–608.

———. 1912: "Epiphany," Hastings Encyclopaedia of Religion and Ethics, V. 330–332.

———. 1937: *The Christmas Festival: An Address Delivered by Kirsopp Lake at the Pierpont Morgan Library 15 December 1935*, New York.

Lambrechts, P. 1962: *Attis, van Herdersknaap tot God*, Brussels.

———. 1967: *Attis en het Feest der Hilariën*, Amsterdam.

Lanciani, R. 1897: *The Ruins and Excavations of Ancient Rome*, New York.

Lane, E. N. (editor). 1996: *Cybele, Attis, and Related Cults: Essays in Memory of M. J. Vermaseren*, Leiden.

Lane Fox, R. 1986: *Pagans and Christians*, New York.

Latte, K. 1967: *Römische Religionsgeschichte*, Second Edition, Munich.

Lazenby, J. F. 1996: *The First Punic War: A Military History*, Stanford CA.

Le Gall, J. 1953: *Recherches sur le Culte du Tibre, Publ. de l'Inst. d'Art et d'Archéol. de l'Université de Paris* II.

Leppin, H. 2003: *Theodosius der Grosse: Auf dem Weg zum christlichen Imperium*, Darmstadt.

Levick, B. 1967: *Roman Colonization in Southern Asia Minor*, Oxford.

———. 2007: *Julia Domna: Syrian Empress*, London.

Liebeschuetz, J. H. W. G. 1979: *Continuity and Change in Roman Religion*, Oxford.

LIMC. 1981–1996: *Lexicon Iconographicum Mythologiae Classicae*, Eight Volumes, Zürich.

Linderski, J. 1995: *Roman Questions: Selected Papers 1958–1993*, Heidelberger althistorische Beiträge und epigraphische Studien 20, Stuttgart.

Littlewood, R. J. 2006: *A Commentary on Ovid: Fasti, Book VI*, Oxford.

Loicq, J. 1964: "Mamurius Veturius et l'ancienne Représentation Italique de l'Année," *Hommages à Jean Bayet*, ed. M. Renard and R. Schilling, Collection Latomus 70, Brussels 401–426.

Lott, J. B. 2004: *The Neighborhoods of Augustan Rome*, Cambridge.

Maas, M. 1992: *John Lydus and the Roman Past: Antiquarianism and Politics in the Age of Justinian*, London and New York.

MacBain, B. 1982: *Prodigy and Expiation: A Study in Religion and Politics in Republican Rome*, Collection Latomus 177, Brussels.

MacMullen, R. 1968: "Constantine and the Miraculous," *GRBS* 9 81–96.
———. 1969: *Constantine*, New York.
———. 1981: *Paganism in the Roman Empire*, New Haven CT.
———. 1984: *Christianizing the Roman Empire (A.D. 100–400)*, New Haven CT.
———. 1993: "The Un-Romanized in Rome," *Diasporas in Antiquity*, Brown Judaic Studies 2.88, ed. S. J. D. Cohen and E. S. French, Atlanta, 47–64 GA.
———. 1997: *Christianity and Paganism in the Fourth to Eighth Centuries*, New Haven CT.
Maddoli, G. 1971: "Il Rito degli Argei e le Origini del Culto di Hera a Roma," *PdP* 26 153–166.
Magdelain, A. 1962: "Cinqu Jours épagomènes à Rome?" *REL* 40 201–227.
Mannhardt, W. 1875: *Der Baumkultus der Germanen und ihre Nachbarstämme*, Berlin.
———. 1877: *Antike Wald und Feldkulte aus nordeuropäischer Überlieferung*, Berlin.
Martin, L. H. 1987: *Hellenistic Religions: An Introduction*, Oxford.
Maryon, H. 1956: "The Colossus of Rhodes," *JHS* 76 68–86.
Matthews, J. F. 1973: "Symmachus and the Oriental Cults," *JRS* 63 175–195.
Mattingly, H. B., and E. A. Sydenham. 1923: *The Roman Imperial Coinage, Volume I, Augustus to Vitellius*, London.
———. 1926: *The Roman Imperial Coinage, Volume II, Vespasian to Hadrian*, London.
———. 1927: *The Roman Imperial Coinage, Volume V Part 1, Valerian I to the Interregnum*, London.
———. 1930: *The Roman Imperial Coinage, Volume III, Antoninus Pius to Commodus*, London.
———. 1938: *The Roman Imperial Coinage, Volume IV Part 2, Macrinus to Pupienus*, London.
McArthur, A. 1953: *The Evolution of the Christian Calendar*, London.
McDonough, C. M. 1997: "Carna, Proca, and the *Strix* on the Kalends of June," *TAPA* 127 315–344.
Merrill, E. T. 1924: "The Roman Calendar and the Regifugium," *CP* 19 20–39.
Michalowski, K. 1960: *Palmyra*, London.
Michels, A. K. 1949: "'The Calendar of Numa' and the Pre-Julian Calendar," *TAPA* 80 320–346.
———. 1953: "The Topography and Interpretation of the Lupercalia," *TAPA* 84 35–59.
———. 1967: *The Calendar of the Roman Republic*, Princeton NJ.
Millar, F. 1993: *The Roman Near East 31 B.C.–A.D. 337*, Cambridge MA.
Momigliano, A. D. (editor). 1963: *The Conflict Between Paganism and Christianity in the Fourth Century*, Oxford.
Mommsen, T. 1883: *Res Gestae Divi Augusti*, Second Edition, Berlin.
Moore, C. H. 1924: "The Duration of the Efficacy of the Taurobolium," *CP* 19 363–365.
Moretti, L. 1985: "Frammenti vecchi e nuovi del Commentario dei Ludi Secolari del 17 A. C.," *Atti della Pontificia Accademia Romana di Archeologia*, Ser. III. Rendiconti Vol. 55–56 361–379.
Munn, M. 2006: *The Mother of the Gods, Athens, and the Tyranny of Asia*, Berkeley CA.
Murdoch, A. 2003: *The Last Pagan: Julian the Apostate and the Death of the Ancient World*, Suttun UK.
Nagy, B. 1985 (1992): "The Argei Puzzle," *AJAH* 10 1–27.
Nassivera, J. C. 1976: "Ancient Temples to Pagan Goddesses and Early Churches to the Virgin in the City of Rome: A Topographical Survey," *Echos du Monde Classique* 20 41–54.

Neverov, O. 1986: "Nero-Helios," *Pagan Gods and Shrines in the Roman Empire*, ed. M. Henig and A. King, Oxford 189–194.
Nilsson, M. P. 1920: *Primitive Time Reckoning*, Lund.
Nock, A. D. 1952: "The Roman Army and the Roman Religious Year," *HTR* 45 187–252.
Noiville, J. 1936: "Les Origines du *Natalis Invicti*," *REA* 38 144–176.
Norden, E. 1924: *Die Geburt des Kindes*, Leipzig-Berlin.
North, J. A. 1979: "Religious Toleration in Republican Rome," *PCPS* N.S. 25 85–103.
Odahl, C. M. 2004: *Constantine and the Christian Empire*, London.
O'Donnell, J. J. 1977: "Paganus," *Classical Folia* 31 163–169.
———. 1978: "The Career of Virius Nicomachus Flavianus," *Phoenix* 32 129–143.
———. 1979: "The Demise of Paganism," *Traditio* 35 43–88.
Ogilvie, R. M. 1965: *A Commentary on Livy Books 1–5*, Oxford (reprinted with additions and corrections 1970).
Olshausen, E. 1978: "Über die römischen Ackerbrüder: Geschichte eines Kultes," *ANRW* II. 16.1 820–832.
Orlin, E. M. 2010: *Foreign Cults in Rome: Creating a Roman Empire*, Oxford.
Orr, D. G. 1978: "Roman Domestic Religion: The Evidence of the Household Shrines," *ANRW* II. 16.2 1557–1591.
Pais, E. 1906: *Ancient Legends of Roman History*, London.
Palmer, R. E. A. 1970: *The Archaic Community of the Romans*, Cambridge MA.
———. 1974: *Roman Religion and Roman Empire*, Philadelphia PA.
———. 1981: "The Topography and Social History of Rome's Trastevere (Southern Sector)," *Proceedings of the American Philosophical Society* 125 368–397.
———. 1990: *Studies of the Northern Campus Martius in Ancient Rome*, Transactions of the American Philosophical Society 80.2, Philadelphia PA.
Parke, H. W. 1977: *Festivals of the Athenians*, Ithaca NY.
———. 1988: *Sibyls and Sibylline Prophecy in Classical Antiquity*, London and New York.
Parker, H. N. 2004: "Why Were the Vestals Virgins? or The Chastity of Women and the Safety of the Roman State," *AJP* 125 563–601.
Parker, R. 1983: *Miasma: Pollution and Purification in Early Greek Religion*, Oxford.
Pascal, C. B. 1979: "Fire on the Tarentum," *AJP* 100 532–537.
Pasoli, A. 1950: *Acta Fratrum Arvalium*, Bologna.
Peter, H. 1914: *Historicorum Romanorum Reliquiae*, Volume I, Second Edition, Leipzig.
Pietrzykowski, M. 1978: "Die Religionspolitik des Kaisers Elagabal," *ANRW* II. 16.3 1806–1825.
Piganiol, A. 1947: *L'Empire Chretien*, Paris.
———. 1962: "Les Origines d'Hercule," *Hommages à Albert Grenier*, ed. Marcel Renard, Collection Latomus 58, Brussels 1261–1264.
Pighi, G. B. 1965: *De Ludis Saecularibus Populi Romani Quiritium Libri Sex*, Editio Altera, Amsterdam.
Pleket, H. W. 1965: "An Aspect of the Emperor Cult: Imperial Mysteries," *HTR* 58 331–347.
Potter, D. S. 1990: *Prophecy and History in the Crisis of the Roman Empire: A Historical Commentary on the Thirteenth Sibylline Oracle*, Oxford.
Poultney, J. W. 1959: *The Bronze Tables of Iguvium*, Baltimore MD.
Powell, B. B. 1991: *Homer and the Origin of the Greek Alphabet*, Cambridge.
Preger, T. 1901: "Konstantinos-Helios," *Hermes* 36 457–469.
Quin-Schofield, W. 1969: "Sol in the Circus Maximus," *Homages à Marcel Renard*, Collection Latomus 102, ed. J. Bibauw, Brussels, II. 639–649.

Radke, G. 1963: "Beobachtungen zum Römischen Kalender," *RM* 106 313–335.

———. 1979: *Die Götter Altitaliens*, Münster.

———. 1990A: *Fasti Romani: Betrachtungen zur Frühgeschichte des römischen Kalenders*, Münster.

———. 1990B: "Gibt es Antworten auf die Argeerfrage?," *Latomus* 49 5–19.

———. 1993: "Römische Feste im Monat März," *Tyche* 8 129–142.

Ramsay, W. M. 1917: "Studies in the Roman Province Galatia, I. The Homonadeis and the Homonadensian War," *JRS* 7 229–283.

Ramsey, J., and A. L. Licht: 1997: *The Comet of 44 B.C. and Caesar's Funeral Games*, Atlanta GA.

Rasmussen, S. 2003: *Public Portents in Republican Rome*, Rome.

Rebuffat, M. R. 1966: "Les Phéniciens à Rome," *MEFR* 78 7–48.

Reitzenstein, R. 1978: *Hellenistic Mystery Religions: Their Basic Ideas and Significance*, London.

Richard, J. C. 1988: "Historiographie et histoire: L'Expédition des Fabii à la Crémère," *Latomus* 47 526–553.

———. 1989: "L'Affaire du Crémère: Recherches sur l'évolution et le Sens de la tradition," *Latomus* 48 312–325.

Richardson, L., Jr. 1992: *A New Topographical Dictionary of Ancient Rome*, Baltimore MD.

Ridgway, D., and F. R. S. Ridgway (editors). 1979: *Italy Before the Romans: The Iron Age, Orientalizing and Etruscan Periods*, London and New York.

Rives, J. B. 1999: "The Decree of Decius and the Religion of Empire," *JRS* 89 135–154.

Robinson, M. 2011: *A Commentary on Ovid's Fasti Book 2*, Oxford.

Rodgers, B. S. 1980: "Constantine's Pagan Vision," *Byzantion* 50 259–278.

Roller, L. E. 1999: *In Search of God the Mother: The Cult of Anatolian Cybele*, Berkeley CA.

Roscher, W. H. (editor). 1884–1937: *Ausführliches Lexikon der griechischen und römischen Mythologie*, Six Volumes, Leipzig.

Rose, H. J. 1923: "A Greek Taurobolic Inscription from Rome," *JHS* 43 194–196.

———. 1924: *The Roman Questions of Plutarch*, Oxford.

———. 1926: *Primitive Culture in Italy*, London.

———. 1933: "The Cult of Volkanus at Rome," *JRS* 23 46–63.

———. 1934: "Two Roman Rites," *CQ* 28 156–158.

———. 1944: "Manes Exite Paterni," *University of California Publications in Classical Philology* 12, Berkeley 89–93.

Roth, R. 1853: "Über die römischen Säcularspiele," *RM* 8 365–376.

Rotondi, G. 1966: *Leges Publicae Populi Romani*, Milan.

Rüpke, J. 1995: *Kalender und Öffentlichkeit: die Geschichte der Repräsentation und religiösen Qualifikation von Zeit in Rom*, Berlin.

———(editor). 2007: *A Companion to Roman Religion*, Oxford.

Rutter, J. B. 1968: "The Three Phases of the Taurobolium," *Phoenix* 22 226–249.

Salzman, M. R. 1990: *On Roman Time: The Codex-Calendar of 354 and the Rhythms of Urban Life in Late Antiquity*, Berkeley CA.

Samuel, A. E. 1972: *Greek and Roman Chronology: Calendars and Years in Classical Antiquity*, Munich.

Saradi-Mendelovici, H. 1970: "Christian Attitudes Toward Pagan Monuments in Late Antiquity and Their Legacy in Later Byzantine Centuries," *Dumbarton Oaks Papers* 44 47–61.

Scheid, J. 1990: *Romulus et ses Frères: Le Collège des Frères arvales, Modèle du Culte public dans la Rome des Empereurs*, Paris.

Schnegg-Köhler, B. 2002: *Die augusteischen Säkularspiele*, Munich-Leipzig.

Scramuzza, V. M. 1940: *The Emperor Claudius*, Oxford.

Scullard, H. H. 1981: *Festivals and Ceremonies of the Roman Republic*, Ithaca NY.
Seeck, O. and G. Veith. 1913: "Die Schlacht am Frigidus," *Klio* 13451–467.
Seyrig, H. 1944: "Le Rameau Mystique," *AJA* 48 20–25.
———. 1971: "Antiquités Syriennes 95: Le Culte du Soleil en Syrie à l'Époque Romaine," *Syria* 48 337–373.
Sherwin-White, A. N. 1963: *Roman Society and Roman Law in the New Testament*, Oxford.
Showerman, G. 1901: *The Great Mother of the Gods*, Madison WI.
Smith, C. J. 2006: *The Roman Clan: The Gens from Ancient Ideology to Modern Anthropology*, Cambridge.
Smith, R. 1995: *Julian's Gods: Religion and Philosophy in the Thought and Action of Julian the Apostate*, London and New York.
Sopgiu, G. 1975: "Aureliano," *ANRW* II. 2 1039–1061.
Sordi, M. 1986: *The Christians and the Roman Empire*, tr. Annabel Bedini, Norman OK.
Southerland, C. H. V. 1967: *The Roman Imperial Coinage, Volume VI, From Diocletian's Reform (A.D. 294) to the Death of Maximinus (A.D. 313)*, London.
Spaeth, B. S. 1996: *The Roman Goddess Ceres*, Austin TX.
Speidel, M. 1978: *The Religion of Iupiter Dolichenus in the Roman Army*, Leiden.
Staples, A. 1998: *From Good Goddess to Vestal Virgins: Sex and Category in Roman Religion*, New York.
Steck, T. D. 2009: *Cult Places and Cultural Change in Republican Italy: A Contextual Approach to Religious Aspects of Rural Society after the Roman Conquest*, Amsterdam.
Stoneman, R. 1992: *Palmyra and its Empire: Zenobia's Revolt Against Rome*, Ann Arbor MI.
Strittmatter, A. 1942: "Christmas and the Epiphany: Origins and Antecedents," *Thought* 17 600–627.
Strong, E., and N. Joliffe. 1924: "The Underground Basilica near the Porta Maggiore," *JHS* 44 65–111.
Sullivan, R. D. 1977: "The Dynasty of Emesa," *ANRW* II. 8 198–219.
Suolahti, J. 1977: "Claudia Insons: Why was a fine imposed on Claudia Ap. f. in 246 B.C.?," *Arctos* N.S. 11 133–151.
Syme, R. 1934: "Galatia and Pamphylia under Augustus: The Governorships of Piso, Quirinius, and Silvanus," *Klio* 27 122–148.
———. 1972: "The *Lapis Tiburtinus*," *Vestigia: Akten des VI internazionallen Kongresses für griechische und lateinische Epigraphik*, Munich 585–601.
Takacs, S. A. 1995: *Isis and Sarapis in the Roman World*, Leiden.
Taverner, E. 1918: "The Roman Farmer and the Moon," *TAPA* 49 67–82.
Taylor, L. R. 1931: *The Divinity of the Roman Emperor*, Middletown CT.
———. 1933: "Quirinius and the Census of Judaea," *AJP* 54 (1933) 120–133.
———. 1934: "New Light on the History of the Secular Games,@ *AJP* 55 101–120.
———. 1936: "M. Titius and the Syrian Comand," *JRS* 26 161–173.
———. 1960: *The Voting Districts of the Roman Republic*, *Papers and Monographs of the American Academy in Rome* 20, Rome.
———. 1966: *Roman Voting Assemblies from the Hannibalic War to the Dictatorship of Caesar*, Ann Arbor MI.
Teixidor, J. 1979: *The Pantheon of Palmyra*, Leiden.
Thomas, G. 1984: "Magna Mater and Attis," *ANRW* II. 17.3 1500–1535.
Toynbee, J. M. C., and J. B. Ward-Perkins. 1956: *The Shrine of St. Peter and its Ancient Setting*, London.
Troster, M. 2008: *Themes, Character, and Politics in Plutarch's Life of Lucullus: The Construction of a Roman Aristocrat*, Historia Einzelschriften 201, Stuttgart.

Turcan, R. 1985: *Heliogabale et le Sacre du Soleil*, Paris.
———. 1996: *The Cults of the Roman Empire*, London.
———. 2000: *Mithra et le Mithriacisme*, Paris.
Ulansey, D. 1989: *The Origins of the Mithraic Mysteries: Cosmology and Salvation in the Ancient World*, Oxford.
Usener, H. 1905: "Sol Invictus," *RM* 60 466–491.
———. 1911: *Das Weihnachtsfest*, Bonn.
Van Berchem, D. 1959–60: "Hercule Melqart à l'Ara Maxima," *Atti della Pontificia Accademia Romana di Archeologia (Serie III) Rendiconti* 32 61–68.
Van Ooteghem, J. 1959: *Lucius Licinius Lucullus, Academie Royale de Belgique, Classe des Lettres et des Sciences Morales et Politiques*, Tome LIII, Fascicule 4, Brussels.
Van Son, D. W. L. 1963: "The Disturbances in Etruria During the Second Punic War," *Mnemosyne* Ser. 4, 16 267–274.
Vermaseren, M. J. 1956: *Corpus Inscriptionum et Monumentorum Religionis Mithriacae*, Volume I, The Hague.
———. 1960: *Corpus Inscriptionum et Monumentorum Religionis Mithriacae*, Volume II, The Hague.
———. 1966: *The Legends of Attis in Greek and Roman Art*, Leiden.
———. 1977: *Cybele and Attis: The Myth and the Cult*, London.
———. 1977–1989: *Corpus Cultus Cybelae Attidisque*, Seven Volumes, Leiden.
Versnel, H. S. 1980: "The Historical Implications," 95–140 of *Lapis Satricanus: Archeologiche Studien van het Nederlands Instituut te Rome*, Scripta Minora V, The Hague.
Von Heintz, H. 1983: *Sol Invictus in Spätantike und frühes Christentum*, Frankfurt am Main.
Wackernagel, J. 1923–1924: "Dies Ater," *ARW* 22 215–217.
Wagenvoort, H. 1956: *Studies in Roman Literature, Culture, and Religion*, Leiden.
Walde, A., and J. B. Hofmann. 1938–1954: *Lateinisches etymologisches Wörterbuch*, Third Edition, Two Volumes, Heidelberg.
Warren, M. 1901: "On Some Ancient and Modern Etymologies," *TAPA* 32 110–120.
Watson, A. 1999: *Aurelian and the Third Century*, London.
Webster, G. 1986: *Celtic Religion in Roman Britain*, Totowa NJ.
Webster, H. 1916: *Rest Days: A Study in Early Law and Morality*, New York.
Weinstock, S. 1957: "Victor and Invictus," *HTR* 50 211–247.
———. 1971: *Divus Julius*, Oxford.
Welles, C. B. 1934: *Royal Correspondence in the Hellenistic Period*, London.
Whatmough, J. 1932: "The Ancient Calendar in Italy outside Rome," *HSCP* 41 157–179.
Wildfang, R. L. 2006: *Rome's Vestal Virgins: A Study of Rome's Vestal Priestesses in the Late Republic and Early Empire*, London and New York.
Williams, F. 1987: *The Panarion of Epiphanius of Salamis, Book I*, New York.
———. 1994: *The Panarion of Epiphanius of Salamis, Books II-III*, New York.
Wiseman, T. P. 1998: *Roman Drama and Roman History*, Exeter.
Wissowa, G. 1912: *Religion und Kultus der Römer*, Second Edition, Munich.
Wuilleumier, P. 1932: "Tarente et Le Tarentum," *REL* 10 127–146.
———. 1938: "Tarente et Le Tarentum," *REL* 16 139–145.
Wytzes, J. 1977: *Der letzte Kampf des Heidentums in Rom*, Leiden.
Zanker, P. 1988: *The Power of Images in the Age of Augustus*, tr. Alan Shapiro, Ann Arbor MI.

Index

Z